Advances in Animal Science and Zoology

This is a numbered series focused on the latest research on animal science and in the field of zoology.

nova
science publishers

www.novapublishers.com

Advances in Animal Science and Zoology

Advances in Animal Science and Zoology. Volume 19
Owen P. Jenkins (Editor)
2022. ISBN: 978-1-68507-522-4 (Hardcover)
2022. ISBN: 978-1-68507-534-7 (eBook)

Advances in Animal Science and Zoology. Volume 18
Owen P. Jenkins (Editor)
2021. ISBN: 978-1-68507-255-1 (Hardcover)
2021. ISBN: 978-1-68507-298-8 (eBook)

Advances in Animal Science and Zoology. Volume 17
Owen P. Jenkins (Editor)
2021. ISBN: 978-1-53619-838-6 (Hardcover)
2021. ISBN: 978-1-53619-862-1 (eBook)

Advances in Animal Science and Zoology. Volume 16
Owen P. Jenkins (Editor)
2020. ISBN: 978-1-53618-713-7 (Hardcover)
2020. ISBN: 978-1-53618-727-4 (eBook)

Advances in Animal Science and Zoology. Volume 15
Owen P. Jenkins (Editor)
2020. ISBN: 978-1-53618-254-5 (Hardcover)
2020. ISBN: 978-1-53618-283-5 (eBook)

Advances in Animal Science and Zoology. Volume 14
Owen P. Jenkins (Editor)
2019. ISBN: 978-1-53616-489-3 (Hardcover)
2019. ISBN: 978-1-53616-490-9 (eBook)

More information about this series can be found at
https://novapublishers.com/product-category/series/advances-in-animal-science-and-zoology/

Owen P. Jenkins
Editor

Advances in Animal Science and Zoology

Volume 20

nova science publishers
www.novapublishers.com

Library of Congress Cataloging-in-Publication Data

ISBN: 979-8-88697-199-6
ISSN: 2162-0962

Published by Nova Science Publishers, Inc. † New York

Contents

Preface

This volume includes six chapters that present updates in animal science and zoology. Chapter One examines the use of zebrafish model system to investigate the pathophysiology of schizophrenia. Chapter Two consolidates facts about melatonin and its involvement in the regulation of fish populations. Chapter Three explores the interaction between the native aquatic gastropod, *Valvata humeralis*, and invasive aquatic gastropod, *Valvata humeralis*, along a biotic gradient of density in a system where it was suspected that an interaction shift was occurring. Chapter Four explores competition and facilitation between these two gastropods. Chapter Five analyzes the role of superoxide dismutase in reptiles under toxicity contexts. Lastly, Chapter Six analyzes and assesses the corpses of three English Bulldog puppies that had anasarca, a fetal malformation that presents generalized edema.

Chapter 1 - Schizophrenia is a debilitating mental illness that affects approximately 1% of population worldwide. In recent years, there has been growing interest in using animal models for schizophrenia research, both to elucidate the complex biological mechanisms underlying the disease and to investigate possible interventions to treat its core symptoms. The zebrafish, although relatively new compared to the rodents, has been an excellent model organism for such purpose. The zebrafish has many favourable characteristics, including high fecundity, small size, and low maintenance, are suitable for high-throughput screening of chemical libraries based on locomotor behavior and gene expressions associated with the disease. The ease of genetic manipulations makes zebrafish an excellent choice in understanding the genetics of schizophrenia. Additionally, the rapid developing nervous system and body translucency during early stages enable the direct visualization of complex mechanisms using novel imaging techniques. In this chapter, the authors will examine the use of zebrafish model system to investigate the pathophysiology of schizophrenia. In particular, the authors will discuss the identification of risk genes associated with schizophrenia through generation of mutant fish, the application of zebrafish in developing different pharmacological models, and the behavioral paradigms to characterise the cognitive and other phenotypes relevant to the disease. The authors finally discuss novel assays and cutting-edge techniques that are potentially useful to unveil the biological circuits underlying the disorder.

Chapter 2 - Melatonin (N-acetyl-5-methoxy-tryptamine), a naturally occurring chronobiotic molecule, is principally produced in the pineal organ in response to ambient light-dark (LD) cycle in a rhythmic fashion. Additionally, it is also synthesized from various extra-pineal sources in fish like brain, retina, gut, liver, ovary etc. This indoleamine is bio-synthesized from aromatic amino acid L-Tryptophan through four enzyme-catalyzed reactions, and arylalkylamine N-acetyltransferase (AANAT), converts serotonin to N-acetyl serotonin, is the rate-limiting enzyme of this pathway. Presence of this ~23 kDa key regulator enzyme in fish pineal, retina, gut, liver is studied elaborately but till date the data on endogenous synthesis of melatonin in reproductive organs are not so sufficient. In last few years, many researchers explored the regulatory pathway of melatonergic system in male and female reproduction and they also hypothesized that this hormone may be applied in aquaculture system for better production of fish to fulfill its demand as table food. Melatonin not only performs a multitude of regulatory activities in seasonal reproduction, it has also a wide range of pleiotropic actions to control the other physiological functions in aquaculture like digestive system, immune system, etc. This pineal hormone acts as a primary synchronizer between environment and the neuroendocrine system for the regulation of seasonal reproduction in vertebrates. However, seasonal breeders rely on their environment to regulate various biological rhythms that control physiological processes such as gonadal maturity, and melatonin, aids in the regulation of this progressions to ensure efficacious reproduction and maximum survival of the offspring. The annual gonadal cycle in majority of seasonally reproducing fish consists of: i) preparatory phase, in which germ cells undergoes gametogenic recrudescence; ii) a long pre-spawning phase, where cells undergo growth and development; iii) a brief spawning phase during which gonadal cells complete their maturation; and iv) post-spawning phase where regression of cells takes place. The reproductive phases in an annual gonadal cycle appear to be markedly under the control of an endogenous clock framework that offers vital information about the "time of year" and changing environmental signals in a species-specific way. All the 'clock-controlled genes' (CCGs) are primarily regulated by the endogenous circadian clock, which is located in the suprachiasmatic nucleus (SCN) of the hypothalamus and in other peripheral organs as well. Expression of aanat gene is a crucial switch in melatonin production and is regulated as a CCG. Clock genes and a complex 'transcriptional and translational feedback loop' (TTFL) govern reproductive cycle, but the involvement of melatonin and the mechanism by which it controls rhythmic molecular events, remain a mystery. Melatonin regulates gonadal development via hypothalamus-hypophyseal-gonadal (HPG) axis in a specific reproductive phase dependent manner. Substantial data gathered from a number of carefully controlled studies employing exogenous melatonin administration and/or altered photoperiods (long- and short photoperiods) to several fish species, especially in carp, concluded that it has stimulatory or inhibitory or no effects at all on the functions of gonads depending on the reproductive status of concerned fish. Recent studies on carp also demonstrated that ovarian melatonin concentration exhibited seasonal variations with titer reaching a peak during the spawning season and then considerably declining all through the post-spawning phase of the annual reproductive cycle. Melatonin may potentially have a function in regulating vitellogenesis in liver, and the incorporation of vitellogenin protein in oocytes (a marker of advancement of germ cells growth and maturation) as well, according to the evidences. Another important contribution from the study on carp ovary clearly suggested that melatonin accelerates the maturation of ovaprim-(synthetic GnRH and domperidone) induced oocytes by activating the structural components of maturation promoting factor (MPF) i.e., cyclin B and Cdc2; and this study also hypothesized that the hormonal combination may be used as "biological tool" in aquaculture for successful spawning. However, detection of high affinity melatonin binding sites (~37 kDa MT1 receptor protein) in carp oocytes argued in favor of receptor mediated action of melatonin in reproduction. Interestingly, due to lipophilic nature of melatonin, it also permeates the plasma membrane in living cells and acts as a potent antioxidant in a receptor independent pathway. It aids in the diminution of oxidative stress by scavenging oxygen centered free radicals directly and also by increasing the activities of several antioxidative enzymes and minimize the ovulatory oxidative stress during spawning phase in fish. A better knowledge on the reproductive physiology of seasonal breeders such as fish may be gained by serious attention on the intricate network of clock genes and their associative interaction with melatonergic system on the regulation of these events. However, despite a great deal of research in the past few decades and current studies, involvement of melatonin in fish reproduction, especially in male, and its potential application in aquaculture is still an enigma. The purpose of this chapter is to consolidate the required facts about melatonin and its involvement in the regulation of fish reproduction in order to provide relevant information for subsequent investigations.

Chapter 3 - Much research has been devoted to understanding how interactions shift along abiotic gradients of stress, but less is known about biotic gradients and how they can affect the strength and intensity of interactions between species. The authors explored the interaction between the native aquatic gastropod, *Valvata humeralis*, and invasive aquatic gastropod, *Valvata humeralis*, along a biotic gradient of density in a system where the authors suspected that an interaction shift was occurring. The authors experimentally quantified the intra- and interspecific net effects of these species on each other and tested the results in a field setting by comparing them with field surveys. *Valvata humeralis* growth was significantly greater in treatments containing *V. humeralis* and *P. antipodarum* at 400/m2 each (total gastropod density 800/m2) when compared to growth of *V. humeralis* alone at 400/m2. The facilitative growth effect disappeared at higher snail densities indicating it was density specific. Field surveys indicated that competition between the species at medium densities did not negatively affect *V. humeralis* growth, but did not show the same facilitation seen in the experiment. Experimental and field results combined suggest that the shifting interaction between species is intertwined with other biotic and abiotic factors in the field. The authors' results suggest that interactions between a native and an invasive species can shift along a biotic density gradient potentially influencing invasive species impact.

Chapter 4 - In species that interact through both competition and facilitation, net interaction strength can change from negative to positive along abiotic or biotic gradients as the dominance of competition and facilitation change. The authors explored competition and facilitation between the aquatic gastropods *Valvata humeralis* (a native) and *Valvata humeralis* (an invasive). To assess potential nutrient facilitation, the authors measured C:N and C:P ratios of *V. humeralis* and *P. antipodarum* body content, excretion, egestion, and field epilithon. To assess competition, the authors measured the effects of intraspecific and interspecific density and food availability on *V. humeralis* growth and mortality with a laboratory growth experiment. *Valvata humeralis* juveniles had a lower ratio of body C:P than *P. antipodarum* adults and both ages of *V. humeralis* indicating that *P. antipodarum* juveniles have a higher phosphorus demand. The authors did not detect a difference between phosphorus excretion and egestion rates of *P. antipodarum* and *V. humeralis* indicating that facilitation by nutrient enhancement was unlikely. In the laboratory growth experiment, relative growth of *V. humeralis* during week one decreased with increasing density regardless of food quantity indicating that interference was the competitive mechanism. However, growth of *V.*

humeralis was increased in high density interspecific treatments during the second week suggesting ontogeny influences the interaction between species. The authors hypothesize that the facilitative effect of *P. antipodarum* on *V. humeralis* was caused by increased access to food and potentially access to food pre-conditioning by *P. antipodarum*. The authors' results provide insight into how interactions between native and invasive species can change with density and ontogeny.

Chapter 5 - Reptiles are important models in evolutionary history and several species are of great ecological, sociological and commercial importance. They are transcendental in natural ecosystems, as prey, predators, herbivores, seed dispersers and as bioindicators of environmental health. In recent decades, the role of environmental pollution as a possible reason for the decline of reptile populations has prompted researchers to evaluate the effects of pesticides on these vertebrate groups. Given the expansion of agricultural frontiers, many reptile populations live now in habitat subjected to high pollution pressure, and due to their biological characteristics, they may be exposed to pollutants at all stages of their lives. As a consequence, they appear worldwide as excellent model species of environmental contamination by different xenobiotics. Environmental pollutants are known to induce oxidative damage to biomolecules, such as proteins, lipids, and nucleic acids, due to the production of reactive oxygen species (ROS). ROS are highly reactive and potentially deleterious to biological systems. To minimize the effects that ROS can generate, aerobic organisms have developed enzymatic and non-enzymatic antioxidant defenses. Enzymatic defenses, such as superoxide dismutase (SOD), catalase (CAT) and glutathione peroxidase (GPx), protect the organism by directly eliminating superoxide radicals and hydrogen peroxide, converting them into less reactive species. Oxidative stress occurs as a result of an imbalance between ROS and antioxidant mechanisms, leading to a variety of physiological and biochemical changes that cause cell deterioration and death. This occurs due to an excessive accumulation of ROS, decrease in antioxidant enzymes, or both. Specifically, SOD is the first line of defense and the most powerful antioxidant in the cell. It catalyzes the dismutation of superoxide anion (O2•-) into hydrogen peroxide (H2O2) and oxygen (O2), making potentially harmful O2•- less dangerous. Four isoforms of SOD enzyme have been described, according to their metallic cofactor; Fe-SOD, Mn-SOD, Ni-SOD and Cu/Zn-SOD, which have distinct subcellular locations. SOD has demonstrated to be sensible to environmental contamination and several studies carried out in reptiles have shown that exposure to xenobiotics produces inhibition or overproduction of SOD or altered sod gene expression in different tissues. Considering the importance of reptiles in the ecosystem and the action of SOD against environmental pollutants, the aim of this chapter is to present the current knowledge into the structure-activity relationship and mechanism of action of SOD in a reptile species, and its importance as a biomarker for environmental contamination in sentinel species.

Chapter 6 - Anasarca is a pathology in English Bulldog dogs, which is a fetal malformation in puppies that presents generalized edema. The greater accumulation of fluid occurs at the subcutaneous level and causes the loss of the normal morphology of the canine; produces dystocic deliveries, in some cases the pups are born alive but their viability is unlikely. The objective of the present work was to analyze and assess by pathological and histological studies the corpses of 3 English Bulldog puppies that were referred to CIESA for diagnosis, a necropsy of the specimens was performed, and samples were collected for histopathological study of most of the tissues, which were preserved in 10% buffered formaldehyde, and prepared with the H&E stain. Macro and microscopic photographic evidence of the lesions was gathered. In the anatomopathological study. In general, the alterations observed in the 3 specimens were: an increase in size with a cylindrical body and soft at palpation, when making the primary cut, ascites was found as the main finding, when approached with a cut. Secondary, hydrothorax, pulmonary hypoplasia in the thoracic cavity was found. While ascites in the abdominal cavity, pulmonary hypoplasia, generalized and hepatic congestion, congestion and intestinal mucoid content was also observed. The histopathological study, in the lung revealed: alveolar collapse, loss of the structure of the blood vessels, the wall of the blood vessels presented different degrees of injury from those that present hydropic degeneration to necrosis of smooth muscle cells, endothelial cell hyperplasia; however, the bronchi did not show defined morphology and structures compatible with hyaline cartilage were not identified, as well as peribronchial and peribronchiolar lymphoid aggregates were not identified. The heart presented undefined myocytes, a scarce amount of trabeculae was observed, also the direction of the cardiac fibers was found in different directions, differing from the normal histological pattern. The results found in the microscopic study indicate that the tissues of the different organs are not fully mature, thus denoting their incompatibility with life. This pathology is very frequently associated with cleft palate. The diagnostic method to be used is ultrasound, a tool that would more accurately indicate the excess fluid in the fetus.

Chapter 1

Zebrafish Models of Schizophrenia: Gaining Insights into the Link from Genes to Brain to Behavior

S. Y. Mok*, PhD
and P. Y. Tang, PhD
Department of Mechatronics and Biomedical Engineering,
Universiti Tunku Abdul Rahman, Kajang, Malaysia

Abstract

Schizophrenia is a debilitating mental illness that affects approximately 1% of population worldwide. In recent years, there has been growing interest in using animal models for schizophrenia research, both to elucidate the complex biological mechanisms underlying the disease and to investigate possible interventions to treat its core symptoms. The zebrafish, although relatively new compared to the rodents, has been an excellent model organism for such purpose. The zebrafish has many favourable characteristics, including high fecundity, small size, and low maintenance, are suitable for high-throughput screening of chemical libraries based on locomotor behavior and gene expressions associated with the disease. The ease of genetic manipulations makes zebrafish an excellent choice in understanding the genetics of schizophrenia. Additionally, the rapid developing nervous system and body translucency during early stages enable the direct visualization of complex mechanisms using novel imaging techniques. In this chapter, we will examine the use of zebrafish model system to investigate the pathophysiology of schizophrenia. In particular, we will discuss the

* Corresponding Author's Email: moksy@utar.edu.my.

In: Advances in Animal Science and Zoology. Volume 20
Editor: Owen P. Jenkins
ISBN: 979-8-88697-199-6

identification of risk genes associated with schizophrenia through generation of mutant fish, the application of zebrafish in developing different pharmacological models, and the behavioral paradigms to characterise the cognitive and other phenotypes relevant to the disease. We finally discuss novel assays and cutting-edge techniques that are potentially useful to unveil the biological circuits underlying the disorder.

Keywords: zebrafish, schizophrenia, gene, neurotransmitter system, behavior

Introduction

Schizophrenia is a complex, chronic mental disorder. The prevalence of schizophrenia is about 1% of the worldwide adult (Bhugra, 2005). This disorder is characterised by a set of positive, negative, and cognitive symptoms (Horacek et al., 2006). Positive symptoms include delusions and disorganisation (Amador and Gorman, 1998). Negative symptoms are constituted by blunted affect, reduced speech output, absence of motivation, social withdrawal, and reduced attention (Blanchard et al., 2011). Cognitive deficits refer to impairment in executive functions, visuo-spatial skills, sustained attention, and poor working memory (Mutsatsa et al., 2006; Ritsner and Blumenkrantz, 2007).

The schizophrenia symptoms are associated with the impairments of neuron volume and number, synaptic connectivity, neurochemical balance, and sensory gating (Olabi et al., 2011). However, the exact neurobiological mechanisms leading to these symptoms still remain unclear and the path to effective antipsychotics development is fraught with hurdles. Therefore, the identification of validated and reliable biomarkers underlying neurobiological mechanisms, and genetic and environmental factors of psychopathology are essential to aid the discovery of efficacious treatments (Caspi and Moffitt, 2006; Nestler, 2013).

Modelling human conditions in animal models enables a high predictive validity of potential therapeutic targets and their underlying molecular interactions. Currently, rodents are the most commonly used models (Meyer and Feldon, 2010), as well as other systems such as rabbits (Esteves et al., 2018), invertebrates such as *Caenorhabditis elegans* (Nass and Hamza, 2007) and *Drosophila melanogaster* (Tolwinski, 2017).

Recently, zebrafish (*Danio rerio*) has become a popular tool to study human diseases (Barut and Zon, 2000), including schizophrenia (Meyer and Feldon, 2010). Zebrafish has physiological and anatomical characteristics of vertebrates and the features of invertebrates, including its transparency and the ease of investigating particular biological phenomena. The advantages of using zebrafish arise from its high fecundity, external fertilization, short generation time, the translucent embryo which allows live imaging, fully sequenced genome, and the ease of genome manipulation (Teame et al., 2019).

In this chapter, we will focus on zebrafish models of the following aspects: genetics, neurotransmission systems, and behaviors of schizophrenia. We will focus on various zebrafish model systems for exploring pathologic conditions associated with schizophrenia risk genes. We will also highlight the basic mechanisms of the major neurotransmission systems and the effects of agonists/antagonists that lead to schizophrenia-like phenotypes. In addition, we will summarize the key behaviors of zebrafish for modelling schizophrenia. Finally, we will discuss current and future advances in imaging of zebrafish behaviors, neural development and activities, leading to a greater understanding of the pathophysiology of schizophrenia.

Zebrafish Schizophrenia Risk Gene Models

It is widely accepted that schizophrenia arises from several genetic and environmental factors (van Os et al., 2008). The risk of developing schizophrenia is approximated at 2% for third-degree relatives, 9% for first-degree relatives, 27% for children of two affected parents, and 50% for monozygotic twins (Ng et al., 2009). Although the heritability of schizophrenia is high (~0.8) (Sullivan, Kendler, and Neale, 2003), its nature goes beyond the Mandelian rules. Genetic polymorphisms with low penetrance and rare genetic variants of high risk, account for an overall risk of schizophrenia (Rujescu, 2017). However, the genetic epidemiology of schizophrenia remains unresolved. The association between one or a few genes with schizophrenia identified by genome-wide linkage scans were lacked of genome-wide significance (Ng et al., 2009).

Zebrafish Genome

The haploid zebrafish genome contains 25 *chromosomes, with an approximate size of two gigabases. The common ancestor of teleost* underwent an additional round of whole-genome duplication called the teleost-specific genome duplication (Meyer and Schartl, 1999). Consequently, zebrafish have a higher number of species-specific genes in their genome than other vertebrates (Collins et al., 2012). Some of these gene duplications remain today (Postlethwait et al., 1999) while some have a new function or no longer expressed in the same tissues as the original genes (Force et al., 1999).

Certain zebrafish genes that are duplicated resulting in at least 70% of the 20,479 human protein-coding genes have at least one zebrafish orthologue. Additionally, 82% of the morbid genes listed in the Online Mendelian Inheritance in Man (OMIM) have at least one zebrafish orthologue. Based on genome-wide association studies (GWAS), at least one zebrafish orthologue for 76% of the 4,023 human genes (Howe et al., 2012).

Most of the mutations in the two orthologues have been occurred in regulatory regions. Both orthologues show similar gene expression patterns to the human orthologue (Chiang et al., 1996; Schauerte et al., 1998). Other than protein coding genes, 4,5556 ribonucleic acid (RNA) genes (Howe et al., 2013), including microRNAs (Nachtigall, Dias, and Pinhal, 2014) were described in zebrafish. Based on these data, the so-called 'reference genome', a high-quality sequence assembly of the zebrafish genome was generated. As a result, zebrafish provides a venue to understand human disease-related genes (Phillips and Westerfield, 2014; White, 2015), including schizophrenia susceptibility genes (Howe et al., 2013). However, due to the gene duplication, certain changes in gene expression and loss-of-function phenotypes caused by toxicity effect or mutation could be masked out (Hill et al., 2005; Gawel et al., 2019). Thus, it may be difficult to draw a conclusion from comparisons with zebrafish and human.

Generation of Mutation in Zebrafish

Zebrafish was firstly used as a genetic model by George Streisinger in 1972. Later, the success of the induction of specific locus (*gol-1*) *germ line mutations* by *gamma-rays* (Walker and Streisinger, 1983) gave rise to new insights into large-scale mutagenesis project in zebrafish. Using ethyl-nitroso-urea (ENU) to mutagenize the male zebrafish, a total of 4264 mutants were obtained (Haffter et al., 1996). Nowadays, the zebrafish is a prominent *in vivo* genetic screening model in neuroscience (Tang et al., 2020). Many mutant and transgenic strains of zebrafish are obtainable through the Zebrafish International Resource Center in Eugene, Oregon (http://zebrafish.org/home/guide.php) (Sprague et al., 2003) and other zebrafish research laboratories.

The most common method for transgenic generation in zebrafish is the use of antisense morpholino oligonucleotides (MOs). After microinjection of embryos, these chemically modified oligonucleotides effectively knock down genes for about 5 days by binding to target messenger RNA (mRNA) and resulting in inhibition of translation (Bill et al., 2009). However, this temporary knockdown limits the investigation of gene function to early stages (Eisen and Smith, 2008). In addition, MOs tend to induce off-target which may lead to nonspecific phenotypes (Kok et al., 2015; Lawson, 2016).

Chemical-mediated random mutagenesis in embryonic zebrafish utilizes ENU to induce random mutations. The mutations are identified using targeting induced local lesions in genomes (TILLING) (Wienholds et al., 2002, 2003; Draper et al., 2004; Moens et al., 2008). The limitation of this approach is the laborious identification of mutated genes by screening zebrafish libraries. Recent advances such as next-generation sequencing further improved the efficiency of this approach (Kettleborough et al., 2013; Varshney et al., 2013; Pan et al., 2015). Compared with chemical mutagenesis, insertional mutagenesis has more favourable features, such as rapid identification of the mutated gene, robustness, and cost-effectiveness in identification of insertional elements based on high-throughput sequencing. In this approach, exogenous deoxyribonucleic acid (DNA) from retroviruses or transposons was used as mutagen and inserted into the genome (Amsterdam, Varshney, and Burgess, 2011; Chen et al., 2013). However, none of a transposon or retroviral vector is as efficient in mutagenizing the genome as ENU (Sakai, Ijaz, and Hoffman, 2018).

Targeted mutagenesis approaches utilize a nonspecific DNA nuclease to bind a DNA binding protein that recognized a specific DNA sequence, to target specific genomic loci in the zebrafish genome. Zinc finger nucleases (ZFN) and transcription activator-like effector nucleases (TALEN) are chimeric fusion proteins designed to bind to a target site within a gene and induce damaging, heritable mutations in a gene of interest in zebrafish (Doyon et al., 2008; Sander et al., 2011). However, these two methods were replaced by clustered regularly interspaced short palindromic repeats (CRISPRs), due to their challenges in predicting the efficiency of gene disruption and identifying a target site within an early exon of a gene of interest (Sakai, Ijaz, and Hoffman, 2018).

Recent studies reported the CRISPR/Cas is a powerful and simple tool for targeted mutagenesis (Cong et al., 2013; Mali et al., 2013). To generate zebrafish mutants, a progammable single guide RNA (sgRNA) is introduced into zebrafish embryos along with cas9 endonuclease mRNA to induce *in vivo* gene targeting. A double stranded break is cleaved by sgRNA-directed Cas9. The repair which relies on the non-homologous end joining (NHEJ) leads to insertion-deletion mutations, and, therefore, the inactivation of a gene of interest (Hwang et al., 2013). The efficacy of CRISPR/Cas to target multiple loci allows simultaneous inactivation of duplicated genes in zebrafish (Jao, Wente, and Chen, 2013). However, a recent study showed that mismatches up to five bases were able to cause off-target mutations (Fu et al., 2013).

Findings of Schizophrenia Susceptibility Genes

Linkage and association studies have implicated genetic contribution to schizophrenia. Hundreds of gene variants are associated with schizophrenia. The roles of these variants in biological mechanisms need to be understood to elucidate disease aetiology, and eventually inform therapeutic strategies. Disrupted-in-schizophrenia 1 (*disc1*) was firstly identified at the breakpoint of a balanced translocation t(1;11) (q42;q14.3) in a large Scottish pedigree (Millar et al., 2000). Later the involvement of the *disc1* locus in psychiatric disorders was proven in the Finnish population (Chubb et al., 2008; Hennah and Porteous, 2009). *disc1 encodes DISC1, an intracellular scaffold protein with multiple interactors involved in the regulation of neurodevelopmental and synaptic processes* (Brandon and Sawa, 2011; Porteous et al., 2011). Since then, animal models including zebrafish were developed to exhibit the effect of *disc1* on behavioral, neurostructural, and neurochemical features related to schizophrenia (Wood et al., 2009; Johnstone et al., 2011; Tomoda et al., 2016). Here, we discuss findings from studies that used zebrafish to investigate the function of schizophrenia susceptibility genes (Table 1).

Although the zebrafish is an attractive animal model for identifying the genetic contribution to schizophrenia, findings are still limited. With its advantages, we believe that the zebrafish can serve as an alternative model to complement rodents. Based on the available data, it appears that most of the

Table 1. Zebrafish models of schizophrenia susceptibility genes

Gene/s	Method	Strain	Age	Findings	Reference
disc1	Antisense MO designed to specifically inhibit translation of *disc1* mRNA	Wild-type AB zebrafish and olig2:egfp	one-cell embryos	A critical requirement for *disc1* in promoting specification of oligodendrocytes oligodendrocyte lineage in the pathogenesis of schizophrenia *disc1* was required for specification of *olig2*-positive cerebellar neurones	Wood et al., 2009
Neuregulin 1 (*nrg1*)	Antisense MO designed to specifically inhibit translation of *nrg1* mRNA	Wild-type AB zebrafish and *olig2:egfp*	one-cell embryos	*nrg1* was required for specification of *olig2*-positive cerebellar neurones	Wood et al., 2009
microRNA 137 (miR-137)	(i) Direct administration of synthetic miR-137 mimic or MO miR-137 antagonist, and (ii) inducible transgenic expression of miR-137 or transgenic anti- miR137 sponge-RNA expression	wild-type zebrafish	one-cell embryos	miR-137 knockdown reduces touch-sensitivity response, an important characteristic involved in schizophrenia	Giacomotto et al., 2016
Mitogen-activated protein kinase 3 (*mapk3*)	A splice blocking MO targeting exon 2 and human wild-type *kctd13* mRNA were injected into embryos	wild-type zebrafish	embryos at the 1- to 4-cell stage	Overexpression of human *kctd13* (potassium channel tetramerization domain containing 13) which interact with *mapk3*, induced both a decrease in head size and a concomitant decrease in the number of cycling cells in the brain	Gusev et al., 2018

Table 1. (Continued)

Gene/s	Method	Strain	Age	Findings	Reference
SH3 and multiple ankyrin repeat domains 3 (*shank3*)	Knocked down the expression of zebrafish *zshank3* orthologous genes (*zs3.1* and *zs3.2*) by injecting selective antisense MOs into blastocysts	-	-	The resulted in a reduction in size of the head, eyes, and trunk and deficit in touch response	Gauthier et al., 2010
atxn7, *akt3*, *gigyf2*	Homozygous mutants were generated by injection of guide RNAs and Cas9 protein into EK/nacre (*mitfa-/-*) embryos and crossing of the mosaic mutant adults to EK or nacre and the offspring were mated to each other.	EK/nacre	embryos	Displayed decreased prepulse inhibition	Thyme et al., 2019
foxg1, *bcl11b*, *akt3*, *gigyf2*, *znf536*, *rora*	Homozygous mutants were generated by injection of guide RNAs and Cas9 protein into EK/nacre (*mitfa-/-*) embryos and crossing of the mosaic mutant adults to EK or nacre and the offspring were mated to each other.	EK/nacre	embryos	Decreased volume in different brain subregions	Thyme et al., 2019
kmt2e	Homozygous mutants were generated by injection of guide RNAs and Cas9 protein into EK/nacre (*mitfa-/-*) embryos and crossing of the mosaic mutant adults to EK or nacre and the offspring were mated to each other.	EK/nacre	embryos	Had increased brain volume	Thyme et al., 2019

Gene/s	Method	Strain	Age	Findings	Reference
csmd1	Homozygous mutants were generated by injection of guide RNAs and Cas9 protein into EK/nacre (*mitfa* -/-) embryos and crossing of the mosaic mutant adults to EK or nacre and the offspring were mated to each other.	EK/nacre	embryos	Increase in brain activity	Thyme et al., 2019
cacna1c, clcn3	Homozygous mutants were generated by injection of guide RNAs and Cas9 protein into EK/nacre (*mitfa* -/-) embryos and crossing of the mosaic mutant adults to EK or nacre and the offspring were mated to each other.	EK/nacre	embryos	Decrease in brain activity	Thyme et al., 2019
Calcium voltage-gated channel subunit alpha1 D *(cacna1d)*	The *sa17298* mutant line which allele with one point mutation was generated by ENU mutagenesis	-	Adult males	Mutation of *cacna1d* resulted hyperlocomotion, a surrogate phenotype psychosis-like behaviors.	Sanger Institute Zebrafish Mutation Project, 2013; Banono et al., 2021
Solute carrier family 6 member 3 (*slc6a3*)	The *slc6a3* CRISPR sgRNAs were co-injected with Cas9 protein into zebrafish embryos	AB zebrafish	1-cell stage	Brain concentrations of DA and tyrosine hydroxylase-immunoreactive neurons in the midbrain in the *slc6a3−/−* fish are decreased compared with wild-type. *The mutant fish* hover near the tank bottom, with a repetitive digging-like behavior	Wang et al., 2019

Table 1. (Continued)

Gene/s	Method	Strain	Age	Findings	Reference
PDZ domain containing 1 (*pdzk1*)	The guide RNA, Cas9 and stop-codon cassette were co-injected into zebrafish zygotes. These *pdzk1*-KO mutants were crossed with the Tg(*ptf1a:GFP*) line.	*pdzk1*-KO/Tg(*ptf1a:GFP*)		Mutants showed enhanced contrast sensitivity, a visual endophenotype of schizophrenia and autism spectrum disorder.	Xie et al., 2021
Tyrosine hydroxylase 1 (*th1*)	MOs targeting the splice donor sites of exon 2 and exon3 of the zebrafish *th1* gene, were injected into zebrafish embryos.	Wild-type zebrafish	1–2-cell stage	By adulthood, when these mutants were exposed to a novel environment the levels of freezing and immediate positioning in deeper zones were significantly reduced.	Formella et al., 2012
Kinesin family member 17 (*kif17*)	Two antisense MOs targeting exon 9 and start codon were injected into zebrafish embryos.	Wild-type zebrafish		In the mutants, intron between exons 8 and 9 was not spliced, leading to a premature stop codon, but the truncated product is probably nonfunctional.	Tarabeux et al., 2010
Regulator of G protein signaling 4 (*rgs4*)	Two morpholinos against the ATG start codon (-4 to -21) and 5' translation start site (-22 to -47) were introduced into blastomeres.	Tu¨ (wild-type) zebrafish	one- to two-cell stage	*rgs4* knockdown did not affect neuron number and patterning but resulted in locomotion defects and aberrant development of axons. *rgs4* knockdown also attenuated the phosphorylation of another schizophrenia candidate gene *Akt1*	Cheng et al., 2013
Potassium channel tetramerization domain containing 13 (*kctd13*)	Splice-blocking morpholino (MO) against *kctd13* injected into embryos at the 1-to 2-cell stage.	wildtype zebrafish	1-to 2-cell stage	The mutants showed reduction of *kctd13* message and increase in head size.	Golzio et al., 2012

Gene/s	Method	Strain	Age	Findings	Reference
T-SNARE domain containing 1 (*tsnare1)*	Human capped mRNA encoding *TSNARE1* was injected in embryos	wild-type ZDR zebrafish	1–4 cell stage	Overexpression of *tsnare1* resulted in a significant decrease in head size	Fromer et al., 2016
furin	A splice-blocking morpholino for *furin a* targeting the splice site donor region of exon 7 was injected into eggs.	wild-type ZDR zebrafish	1- to 4-cell stage	Suppression of furin a led to a decrease in head size which could be attributed to changes in cell proliferation and/or apoptosis	Fromer et al., 2016
Contactin 4 (*cntn4)*	Human capped mRNA encoding *cntn4* was injected in embryos	wild-type ZDR zebrafish	1–4 cell stage	Overexpression of *cntn4* resulted in a significant decrease in head size	Fromer et al., 2016
Copy-number variations (CNVs) in 16p11.2	Capped mRNA for the nine human genes (*cd19*, *nfatc2ip*, *atxn2l*, *tufm*, *atp2a1*, *rabep2*, *spns1*, *lat*, and *sh2b1)* encompassed within the 16p11.2 220 kb BP2-BP3 interval and modeled the duplication by expressing each individual transcript in zebrafish embryos.	wild-type zebrafish	1- to 2-cell stage	Overexpression of *lat* induced a reduction of brain proliferating cells and concomitant microcephaly	Loviglio et al., 2017
Linker for activation of T cells (*lat*)	CRISPR gRNA that targets exon 2 of *lat, with Cas9 protein* into zebrafish embryos	wild-type zebrafish	1-cell stage	Suppression of *lat* induced an increase in the proliferation in the brain of the and macrocephaly.	Loviglio et al., 2017
Nicotinate phosphoribosyl-transferase domain containing 1 (*naprt1*)	A transgene construct, presenting a ubiquitous promoter driving expression of the red fluorescent protein marker and 3 different synthetic microRNAs targeting *naprt1* mRNA was injected into the zebrafish embryos along with transposase to promote its integration into the zebrafish genome	-	1-cell stage	The expression of anti *naprt1* microRNAs triggered the developmental defect holoprosencephaly, in which the brain failed to fully divide into 2 hemispheres.	Periyasamy et al., 2019

schizophrenia susceptibility genes have been investigated using conventional morpholino knockdown technology. Nevertheless, some recent studies have moved forward to apply advent of CRISPR technology, which is particularly essential for confirming the feasibility of generating CRISPR mutants for phenotypes of schizophrenia.

Conserved Schizophreia-Relevant Neurotransmitter Pathways in Zebrafish

Aberrant neurotransmission systems are the main causes of schizophrenia. Dopaminergic, glutamatergic, gamma-aminobutyric acid (GABA)-ergic, serotonergic, cholinergic systems and their interactions are the key mediators implicated in schizophrenia (Benes, 2009). The notion that neurotransmission systems are conserved among vertebrates (Kaslin and Panula, 2001; Sallinen et al., 2009; Maximino and Herculano, 2010; Sundvik and Panula, 2012), zebrafish possess all components of neurotransmission systems, such as transmitters, receptors, transporters, and enzymes needed for transmitter synthesis (Kalueff, Stewart, and Gerlai, 2014; Stewart et al., 2014). Due to their well-characterised systems, both larval and adult zebrafish emerged as a model organism in understanding schizophrenia.

Glutamatergic System

Glutamate system functions as the major excitatory neurotransmitter in the central nervous system (CNS) and serves as a precursor to GABA (Bak, Schousboe, and Waagepetersen, 2006). The levels of extracellular glutamate are tightly regulated (Niciu et al., 2012). Glutamate is synthesized from two primary sources, α-ketoglutarate -a tricarboxylic acid (TCA)-cycle intermediate- by aminotransferase, or from glutamine, catalysed by glutamine synthase, in axon terminals of glutamatergic neurons. It is then uploaded by vesicular glutamate transporters into presynaptic vesicles and released into the presynaptic neuron with the aid of calcium influx. Glutamate is released into synapse and taken up by postsynaptic neuron and glia, or recycled in the presynaptic neuron (Marsman et al., 2013). Dysregulation of this transmission results in increased levels of extrasynaptic glutamate and leads to cellular damage (Léveillé et al., 2008; Xu et al., 2009), which is implicated in the progressive loss of brain tissue in schizophrenia (Marsman et al., 2013). Glutamate also serves as a precursor to GABA (Bak, Schousboe, and Waagepetersen, 2006), and they work together to maintain an excitation/ inhibition balance. N-methyl-D-aspartate (NMDA) receptor hypofunction may trigger the repetitive stereotyped circling, a prominent clinical hallmark of psychosis and schizophrenia (Paoletti, Bellone, and Zhou, 2013; Howes, McCutcheon, and Stone, 2015).

Vesicular glutamate transporters (VGLUT 1, 2, 3, respectively) are responsible for uploading glutamate into synaptic vesicles. VGLUT1 and VGLUT2 are specifically found in glutamatergic neurons (Fremeau et al., 2004), while VGLUT3 is uncharacteristically found in cholinergic, serotonergic, and GABAergic neurons (Fremeau et al., 2002; Gras et al., 2002; Schäfer et al., 2002; Takamori et al., 2002). VGLUT2 starts to develop in Rohon Beard neurons and the mediodorsal domain of the zebrafish spinal cord at 20 to 24 hours post fertilization (hpf). At this early stage, VGLUT1 is expressed in a very limited region of the brain. Its expression in spinal cord in animals only starts at 5 days post fertilization (dpf) (Higashijima, Mandel, and Fetcho, 2004). The glutamate reuptake into glia and neurons is performed by the high affinity excitatory amino acid transporters (EAATs) encoded by the *slc1a* family of genes. In human, five subtypes of EAATs (1-5) have been identified (Lehre et al., 1995; Pow and Barnett, 2000; Furness et al., 2008; Holmseth et al., 2009, 2012) while zebrafish have four of the five EAAT related genes, including three EAAT1 paralogs, EAAT2, EAAT3, and two EAAT5 sequences (Rico et al., 2010).

Glutamate receptors are classified into two categories, metabotropic and ionotropic. Metabotropic receptors are classified into Group I, Group II, and Group III receptors. The human metabotropic receptors have eight subtypes divided into these three groups. In contrast, zebrafish have 12 receptor subtypes divided into three groups (Huang et al., 2012; Haug et al., 2013). Ionotropic receptors include three families; NMDA, α-amino-3-hydroxy-5-methyl-4-isoxazole propionic acid (AMPA) and kainate receptors. The NMDA receptor hypofunction hypothesis has successfully explained the symptoms associated with schizophrenia (Coyle, 2012). Administration of non-competitive NMDA receptor antagonist (dizocilpine, also known as MK-801) can produce a range of schizophrenia-related behavioral phenotypes, such as disrupted social interaction (Benvenutti et al., 2021a), increased circling with no preferred direction (Swain, Sigstad, and Scalzo, 2004), and impaired habituation (Png et al., 2021) in zebrafish.

GABAergic System

GABA is the primary inhibitory neurotransmitter in the adult CNS, synthesized by decarboxylation of glutamate by two isotypes of glutamic acid decarboxylase (GAD), GAD65 and GAD67 (Petroff, 2002; Thompson et al., 2009). GAD65 is highly distributed in axon terminals and synaptic vesicle membranes to regulate GABA synthesis by short-term mechanisms (Soghomonian and Martin, 1998; Tian et al., 1999). GAD67 is distributed throughout GABAergic neurons and accounts for basal GABA synthesis (Hashimoto et al., 2003; Curley et al., 2011) and the majority of GABA synthesis (Asada et al., 1996). A GABA defect which involves decreased expression of transcripts for GAD67 occurs in dorsolateral prefrontal cortex (DLPFC) in layers 1 through 5 of schizophrenia patients (Akbarian et al., 1995; Curley et al., 2011). The two *gad* complementary DNAs from zebrafish are over 84% identical to human *gad65* and *gad76*. Both *gad* genes are expressed in the early zebrafish embryo during the period of axonogenesis. These genes are detected in the telencephalon, midbrain, border regions of the rhombomeres in the rostral hindbrain, and spinal cord. Only *gad67* is detected in the caudal hindbrain (Martin et al., 1998).

There are two distinct classes of GABA receptor, ionotropic $GABA_A$ and metabotropic $GABA_B$. $GABA_A$ receptor is a ligand-gated chloride channel that mediates the major form of fast inhibitory neurotransmission through rapid postsynaptic membrane hyperpolarization (Vithlani, Terunuma, and Moss, 2011). This receptor is formed by a family of 19 subunits encoded by distinct genes: α(1–6), β(1–3), γ(1–3), δ, ε, θ, π, and ρ(1–3) (Korpi et al., 2002; Rudolph and Möhler, 2006; Whiting, 2006). Compared with human, zebrafish genome contains 22 genes encoding for $GABA_A$ receptor subunits (α_1–α_{6b}, β_1–β_4, γ_1–γ , δ, π, ζ, ρ_1–ρ_{3a}) and 7 subunit-like genes (α_2-like, α_3-like, two β_2-like, π-like, ρ_1-like, ρ_3-like) (Cocco et al., 2017). The α_1 subunit is located in the cerebellum (Delgado and Schmachtenberg, 2008) while the subunits α_1, α_3, ρ_1, ρ_1-like, ρ_{2a}, ρ_{2b} are found in the retina (Connaughton et al., 2008).

$GABA_B$ receptor, a G-protein coupled receptor (GPCR) which couples indirectly to K^+ channels, produces slow and prolonged inhibitory signals (Bettler et al., 2004). $GABA_B$ receptor consists of two subunits: $GABA_{B1}$ and $GABA_{B2}$ (Kaupmann et al., 1998; Marshall et al., 1999).. In zebrafish $GABA_B$ receptors are found in the cerebellum. Of the three zebrafish genes for the $GABA_B$ receptor, two of them, *gabbr1a* and *gabbr1b*, are homologous to the human gene for subunit $GABA_{B1}$, while *gabbr2* is homologous to human $GABA_{B2}$ (Delgado and Schmachten-berg, 2008).

An association between schizophrenia and single nucleotide polymorphisms (SNPs) in the GABRB2 gene, which encodes the $GABA_A$ receptor β2 subunit was observed in the Chinese Han, German, Japanese, and Portuguese populations (Lo et al., 2004, 2007; Zhao et al., 2006). The alternatively spliced *gabbr2* isoforms resulted in different electrophysiological consequences that may play a key role in the pathogenesis of schizophrenia (Zhao et al., 2006, 2009). In addition, the decrease of *gabbr2* expression in brains is genotype- and age-dependent in schizophrenia patients (Zhao et al., 2012). Furthermore, hypermethylation in the neighbourhood of disease-associated SNPs in schizophrenia patients indicates this may be a plausible underlying mechanism of schizophrenia (Pun et al., 2011). In a zebrafish model, an aberrant DNA methylation pattern at the promoter of *gabrb2* was found to induce schizophrenia-like symptoms (Wang et al., 2016), and the study further supported the previous findings.

In the GABAergic neurotransmission, the main function of GABA transporters is to regulate the level of GABA at the synaptic cleft, by binding the extracellular GABA and transporting it into the cytoplasm (Zhou and Danbolt, 2013; Schousboe et al., 2014). In human, there are four types of GABA transporters: GAT-1, BGT-1, GAT-2, and GAT-3 (Scimemi, 2014). All four proteins belong to one family of sodium-dependent membrane transporters, the solute carrier 6 (SLC6) family (Kristensen et al., 2011). In schizophrenia patients, the downregulation of GAT-1 was observed in prefrontal cortex (Volk et al., 2001), limbic system (Hashimoto et al., 2008), and cerebellum (Bullock et al., 2008), whereas an increase of GAT-3 density was detected within DLPFC (Schleimer et al., 2004), suggesting changes in GABA neurotransmission may contribute to the etiology of schizophrenia.

In zebrafish, four transporters, *slc6a1a*, *slc6a1b*, *slc6a11a*, and *slc6a13* were identified. *slc6a1a* which is highly expressed in whole brains and brain regions is homologous to human GAT-1. *slc6a1b* is less abundant than *slc6a1a*, while *slc6a11a* is scarcely detected in whole brains and brain regions. The transporter *slc6a13* is less detected than *slc6a1a* and *slc6a1b* in all brain regions except the diencephalon (Cocco et al., 2017). Nonetheless, given the potential role of GABA transporter in neurotransmission, the functional significance of GABA transporter associated with schizophrenia remains unknown in the zebrafish model.

Dopaminergic System

Dopamine (DA) is a predominant catecholamine. In mammals, DA is produced in the midbrain nuclei; mainly ventral tegmental area (VTA) and substantia nigra pars compacta (Poulin et al., 2018). The mesocortical pathway originates in the VTA and action potentials travel to the prefrontal cortex. The mesolimbic pathway projects action potentials from the VTA to nucleus accumbens while the DA neurons in the nigrostriatal pathway begin in the substantia nigra pars compacta and project to the dorsal striatum (Rink and Wullimann, 2001). The dopaminergic system is distinct between zebrafish and mammals. Compared with mammals, zebrafish does not have a dopaminergic system in midbrain. The dopaminergic cells are localized in the diencephalon (DC), hindbrain and subpallium (Rink and Wullimann, 2001, 2002). The ventral DC is considered as the substantia nigra. The dopaminergic neurons in ventral DC can be divided into seven cell populations, which are DC1, DC2, DC3, DC4, DC5, DC6, and DC7 (Blechman et al., 2007).

DA is synthesized in terminal of dopaminergic neurons, from tyrosine which is transported across the blood brain barrier (Nakashima et al., 1999). Tyrosine is then converted into L-3,4-dihydroxy-phenylalanine (*L-DOPA*) and form DA. DA is transported from cytosol into the vesicular compartment by the monoamine transporter and is released into the extracellular space. The extracellular DA is taken up into the terminal by the DA transporters (DATs) and is removed from the system by uptake into glial cells and the blood and diffusion out of the striatum (Jones et al., 1998). DA acts on two classes of GPCRs. D1-class receptors consists of D1 and D5 which couple to G stimulatory sites and activate cyclic adenosine monophosphate (cAMP) production and phospholipase C activity. D2, D3, and D4 are grouped together as D2-class receptors. These receptors couple to G inhibitory sites, which inhibit the production of cAMP (Beaulieu and Gainetdinov, 2011). Genes encoding zebrafish D2, D3, and D4 DA receptors have been cloned and characterised. Due to the genome duplication, three zebrafish D2 and D4 DA receptor genes have been identified. A single zebrafish orthologue of the mammalian D1 and D3 receptor genes were identified, suggesting the loss of the second orthologues during zebrafish evolution (Boehmler et al. 2004; 2007; Ek et al., 2016).

In a zebrafish model, effects of dopaminergic modulation on protein kinase *Akt*, which is known to be downregulated in the brain of patients with schizophrenia (Emamian et al., 2004), were demonstrated. This D2-dependent reduction in *Akt* activity suppressed motor behavior, GSK3b activity and decreased the size of the GABAergic neuron population throughout the brain (Souza, Romano-Silva, and Tropepe, 2011). The hypoactive effect of clozapine, the preferred drug for treatment-resistant schizophrenia (Tandon and Fleischhacker, 2005), was found to act through D4 in zebrafish larvae (Boehmler et al., 2007). In addition, adult zebrafish exposed to this antipsychotic showed downregulation of *sco1*, *ocd1,* and *arrdc3* which play a role in functional pathways relevant to schizophrenia (Viana et al., 2020). *sco1* which encodes *cytochrome C oxidase assembly protein 1*, is essential for copper homoeostasis (Leary et al., 2007). Schizophrenia patients have been shown to exhibit significantly higher plasma copper concentrations (Vidović et al., 2013). *ocd1* encodes *ornithine decarboxylase 1* that catalyzes the conversion of ornithine into putrescine. As a polyamine, putrescine regulates cell growth and differentiation, metabolic pathways, and cell membrane functions (Pegg, 2006, 2016), and has been suggested to be involved in the pathophysiology of schizophrenia (Fiori and Turecki, 2008). *arrdc3*, encoding *arrestin domain containing 3* regulates cell-surface expression of adrenergic receptors and other GPCRs (Gurevich and Gurevich, 2006; Nabhan et al., 2010; Patwari et al., 2011; Han et al., 2013). These observations explain the high expression of *arrdc* in the prefrontal cortex in schizophrenia (Corley et al., 2016).

DAT (SLC6A3) is a plasma membrane protein selectively expressed in dopaminergic neurons. It regulates DA transmission by reuptake of DA into presynaptic terminals and termination of DA's effect (Shimada et al., 1992; Ciliax et al., 1995, 1999). The rare DAT deficiency is believed to be underdiagnosed (Ng et al., 2014), and polymorphisms in the DAT gene (*slc6a3*) have been associated with treatment-resistant schizophrenia (Bilic et al., 2014). DAT-mutant zebrafish have been used as model to investigate pharmacological responses. In *slc6a3−/−* zebrafish, cellular loss of tyrosine hydroxylase (TH)-immunoreactive neurons in the midbrain and abnormal swimming behavior were observed. These abnormalities were reversed by clozapine (Wang et al., 2019), an antipsychotic for treatment-resistant schizophrenia, which acts on D2 receptors (Seeman, 2014).

Serotonergic System

Serotonin (5-hydroxytryptamine, 5-HT) is one of the major neurotransmitters in the CNS. In human brain, the raphe nuclei in brain stem contains the majority of the serotonergic cells (Charnay and Léger, 2010). In zebrafish, the principal regions of serotonergic system are the raphe nuclei, the preoptic area and the basal hypothalamus (Lillesaar, 2011). The synthesis starts with the conversion of tryptophan to 5-hydroxytryptophan (5-HTP) (Walther and Bader, 2003). 5-HT is then synthesized from 5-HTP and transported from the cytoplasm to synaptic vesicles by the vesicular transporter (Mohammad-Zadeh, Moses, and Gwaltney-Brant, 2008). When 5-HT is released to the synaptic cleft, the serotonin transporter uploads it to the serotonergic neurons. Serotonin is then metabolized by monoamine oxidase to 5-hydroxy-indolecetaldehyde to form 5-hydroxyindoleacetic acid (5-HIAA)(Hensler, 2012). 5-HT plays an essential role in emotional processing and affect regulation (Lucki, 1998; Dayan and Huys, 2008). Thus, 5-HT dysregulation could underlie the negative symptoms and co-morbid depression in schizophrenia (Singh et al., 2010).

5-HT receptors are GPCRs and ligand-gated ion channels (Barnes and Sharp, 1999). The main 5-HT receptors and their subtypes are 5-HT1 (5-HT1A, 5-HT1B, 5-HT1D, 5-HT1E and 5-HT1F), 5-HT2 (5-HT2A, 5-HT2B and 5-HT2C), 5-HT3, 5-HT4, 5-HT5 (5-HT5A, 5-HT5B), 5-HT6, and 5-HT7 (Pithadia and Jain, 2009). While on the contrary, zebrafish express four 5-HT receptors (*htr1aa*, *htr1ab*, *htr1bd*, and *htr2c*). Their expression profiles are in line with the homologs from other vertebrates (Norton, Folchert, and Bally-Cuif, 2008; Schneider et al., 2012). SNPs in 5-HT1A, 5-HT2A, and 5-HT2C genes have been suggested to have potential roles in psychotic symptoms, cognitive disturbances, and drug response in schizophrenia (Masellis et al., 1998; Drago, Ronchi, and Serretti, 2008). Moreover, 5-HT receptors represent promising drug targets for antipsychotics (Meltzer et al., 2003; Abi-Dargham, 2007; Nikiforuk et al., 2016). However, pharmacological effect of antipsychotics targeting 5-HT receptors has not been previously described in zebrafish.

The serotonin transporter, encoded by *slc6a4* gene, belongs to a larger family of sodium chloride dependent transporters (Nelson, 1998; Bayle et al., 2003). It plays a key role in regulating 5-HT transmission which involves synaptic 5-HT concentrations and signalling by rapid reuptake of released 5-HT (Kristensen et al., 2011). 5-HT transporter is considered as a candidate for vulnerability to the two main symptoms of schizophrenia, depression and hallucinations (Breier, 1995; Malhotra et al., 1998; Golimbet et al., 2004). 5-HT from zebrafish has been characterised and compared its pharmacological profile to that of the human serotonin transporter. The overall identity between zebrafish and human 5-HT transporters is 63.9%. Zebrafish 5-HT transporter contains 16 additional amino acids with 12 amino acids in the N-terminal, three amino acids in the C-terminal, and a single extra amino acid in the large extracellular loop 2. It was also found that the zebrafish 5-HT transporter has a higher affinity towards imipramine compared to human 5-HT transporter (Severinsen et al., 2008).

Cholinergic System

Cholinergic neurotransmission plays a crucial role in a variety of CNS functions including the regulation of attention, sleep, and higher-order cognitive processing (Raedler and Tandon, 2006). The systems are anatomically diffuse and highly interactive with other neurotransmitter and neuromodulator systems (Woolf, 1991; Karczmar, 1993). The system network consists: (i) the basal forebrain complex, (ii) the pedunculopontine and laterodorsal tegmental distributed from the brain stem to the thalamus, midbrain and other brain stem regions, and (iii) interneurons in the striatum and the nucleus accumbens (Everitt and Robbins, 1997; Perry et al., 1999).

Cholinergic receptors are arranged into two subtypes. The muscarinic subtype of cholinergic receptors are G-protein-coupled while the nicotinic subtype is a ligand-gated ion channel. Five muscarinic and nine nicotinic receptor genes have been identified (Arneric et al., 1995; Wess et al., 1995; 2003; Eglen and Watson, 1996; Wess, 1996; 2004). There are five subtypes of muscarinic receptors (M_1–M_5) (Caulfield and Birdsall, 1998). The most abundant nicotinic receptors in the brain are of the forms of the heteromeric $\alpha 4\beta 2$ and the homomeric $\alpha 7$ forms (Gotti et al., 2009).

Dysfunction of the cholinergic system, particularly muscarinic and nicotinic receptors have been assigned the role in the pathogenesis of schizophrenia (Dean et al., 1996; Court et al.. 2000). Activation of excitatory cholinergic muscarinic M1 receptors (CHRM1) and inhibitory CHRM4 receptors lead to a behavioral and cognitive impairment, suggesting that these receptors may be involved in the pathophysiology of schizophrenia (Santiago and Potter, 2001). A decreased binding of the $\alpha 7$ receptor to ligand was observed in hippocampi and thalamic nuclei of schizophrenia patients (Freedman et al., 1995; Court et al., 1999). In addition, a reduction in protein expression and mRNA levels of gene encoding the α7 receptor have been reported in individuals with schizophrenia (Guan et al., 2000; Mexal et al., 2010).

The zebrafish has a relative similar cholinergic system to other vertebrates. Neuronal cells containing choline acetyltransferase are distributed in the telencephalon, preoptic region, diencephalon, mesencephalon, isthmic region, perilemniscal nucleus, superior reticular nucleus, and rhombencephalon (Mueller, Vernier, and Wullimann, 2004). Despite the identification of cholinergic neurons in zebrafish CNS, the development of zebrafish model to study the dysfunction of cholinergic neurotransmission in schizophrenia and cholinergic receptors as antipsychotic targets has not been established.

Schizophreia-Relevant Zebrafish Behaviors

Zebrafish exhibit a rich repertoire of behaviors with quantifiable responses. These behaviors are robust and appear to be evolutionarily conserved that they resemble those of mammalian (Kalueff et al., 2013). Many works have been focused in recent years to identify endophenotypes to quantify deficits in diseases and which may be used to assay potential treatments (Orger and de Polavieja, 2017). This section discusses the behaviors of zebrafish that are of potential relevance to the three domains of schizophrenia symptomatology. The recent approaches that are used to study these behaviors, together with their potentials and limitations are also explored.

Prepulse Inhibition

Defined as the reduced response to startle stimuli following presentation of a lower-intensity stimulus (or prepulse), prepulse inhibition (PPI) is an important clinical measure of sensorimotor gating deficits in schizophrenia and a few other diseases (Braff, Geyer, and Swerdlow, 2001). This evolutionarily conserved mechanism is readily observable in humans and in other vertebrates (Geyer, McIlwain, and Paylor, 2002). As in mammals, PPI in zebrafish is sensitive and susceptible to genetic and pharmacological disruptions (Bergeron et al., 2015; reviewed in Gawel et al., 2019; Banono and Esguerra, 2020). Although the neural basis of PPI in zebrafish is relatively primitive compared to mammals, PPI was consistently demonstrated in larvae starting 5-6 dpf in a manner distinctly similar to mammalian PPI, suggesting the relevance of larval zebrafish in studying the basic mechanisms underlying sensorimotor gating (Burgess and Granato, 2007; Langova et al., 2020). Indeed, zebrafish PPI has been a promising endophenotype for genetic analysis of schizophrenia (Burgess and Granato, 2007).

A common method to assay PPI is assessment of startle response to acoustic stimuli of varying intensity and frequency. In zebrafish, startle response is expressed as an escape reflex called C-start that occurs on two distinct time scales (Burgess and Granato, 2007). PPI is typically quantified from rate and latency (Burgess and Granato, 2007; Bhandiwad et al., 2013) and/ or amplitude (Gawel et al., 2019; Joo et al., 2021) of C-start responses. The short-latency C-start, which is often an important measure of PPI (Bergeron et al., 2015) is believed to be driven by two giant Mauthner neurons based on mounting evidence from ablation studies (Burgess and Granato, 2007; Hecker et al., 2020) and *in vivo* calcium imaging (Kohashi and Oda, 2008; Takahashi et al., 2017).

Recently, Thyme et al. (2019) have identified two zebrafish mutants for schizophrenia risk genes, *atxn7* and *sbno1* with decreased PPI from their phenotypic analysis of more than 20 mutants created using CRISPR-Cas9. Notably, the PPI phenotype is dependent on the frequency of acoustic stimuli, such that only selective stimulus frequency resulted in impaired PPI response (Joo et al., 2021). The findings from this study, together with other previous work (Medan and Preuss, 2011; Bergeron et al., 2015) highlight the need for a more careful interpretation of PPI results with regards of acoustic signal properties.

Learning and Memory Deficits

Cognitive deficits are core symptoms of schizophrenia and present as key indicators of functional and treatment outcomes. Zebrafish, although is a simple vertebrate, possess sophisticated sensory, motor, and reward systems making them well-suited to experiments on learning and cognition. One of the most popular experiments in zebrafish being mazes, which have been used for multi-behavioral assessment including associative learning, spatial memory, place preference, and even memory consolidation (Blank et al., 2009; Gerlai, 2010; Cognato et al., 2012; Zhang et al., 2018). Depending on the task parameters, mazes for zebrafish studies can come in various designs and complexities (reviewed in Benvenutti et al., 2021b). In addition, a diverse array of experiments has also been devised to study the different forms of learning, including habituation, attention, and working memory (reviewed in Gerlai, 2016 and Langova et al., 2020), which are frequently impaired in schizophrenia and other neurological diseases.

Although restricted to adult zebrafish in the early stage, larvae have quickly gained their popularity in learning and memory research by the end of 20th century. Not only that the simplest forms of learning (habituation, dishabituation, and sensitization) were demonstrated in larval zebrafish (reviewed in Roberts, Bill, and Glanzman, 2013), but the more complex behaviors such as novel object recognition (Andersson, Ek, and Olsson, 2015) and long-term memories (Hinz et al., 2013) have also been successfully measured. These successes are encouraging, as larval zebrafish are amenable to a large range of molecular and optogenetic tools, are more easily maintained, and ideally lend themselves for high-throughput analysis. Together, they offer greater promise to dissect the molecular and the neural circuits underlying learning and memory in putative schizophrenia models.

To date, while reports on cognition studies in zebrafish are not as comprehensive as in rodents, they have provided valuable insights into modelling different types of learning and memory deficits especially with MK-801 (reviewed in Gawel et al., 2019). It remains to be elucidated whether compounds targeting different neurotransmitter systems and pathways could elicit similar deficits.

One major challenge that impedes the translational value of cognitive research utilizing animal models is the increasingly diverse assays with non-comparable behavioral measures and test results with humans. To facilitate the comparisons, it would be valuable to design and develop cross-species behavioral paradigms. One such work is the novel Y-maze test with a series of two-choice discrimination tasks to measure the performance of working memory and cognitive flexibility (Cleal et al., 2021). This assay was shown to be sensitive enough to assess learning in humans and across different animal species including *Drosophila*, zebrafish, and mice, laying a foundation for future translational studies.

Shoaling and Social Preference

Social interactions constitute an important domain of human behaviors. Deficits in social functioning is both a core attribute and a state marker of risk for schizophrenia and other severe brain illness (Porcelli et al., 2019). Although relatively new in neurobehavioral research, the utility of zebrafish in the studies of social phenotypes is rapidly developing due to their highly social nature and robustness in displaying complex social behaviors.

Zebrafish actively form shoals in both natural and laboratory conditions to improve foraging efficiency and predator detection or defence (Eguiraun et al., 2018). This complex behavior of coordinated movements is sensitive to environmental manipulations, and is one of the most popular models to quantify social behaviors (Facciol and Gerlai, 2020). Shoaling can be easily assayed by leaving a group of fish in a tank, where their interactions are tracked based on quantifiable endpoints such as average inter-fish distance (a measure of social cohesion)(Figure 1A), nearest neighbour distance, and furthest neighbour distance (Pham et al., 2012). While it is possible to manually analyze the behavioral parameters based on static images captured from pre-recorded videos, examination of shoaling behavior in recent years has benefited from open-source or commercial computer vision programmes that can automatically and simultaneously track the behavior of multiple fishes (reviewed in Franco-Restrepo, Forero, and Vargas, 2019). These tracking software offer obvious advantages over manual analysis, but nonetheless the identity of individual fish is not maintained. More recently, a deep learning approach has been developed to reliably track the individual activity trajectories of up to 100 fishes (Romero-Ferrero et al., 2019), offer great promise to understand how collective behaviors like shoaling emerged from interactions among individual fish.

Social preference test is another simple yet versatile assay to measure social interaction. Defined as the innate preference to live near conspecifics, this behavior is routinely assayed by observing how an individual fish respond to a social stimulus. A three-compartment tank separated with transparent barriers is often employed for the social preference test (Figure 1B), similar to the three-chamber assay used in rodents. The preference to socialize is tracked from the time the test fish in the middle compartment spent near the compartment where the social stimulus (conspecific) is placed. Depending on the objective of the study, the third compartment can be added with another social stimulus (eg. fish of different strain), a non-social stimulus (eg. food), or left empty, and various 'dummy' social stimuli can be used in place of real fish, such as animated fish image (Gerlai, 2017), 3D-printed fish (Bartolini et al., 2016), and robotic fish (Kopman et al., 2012; Polverino and Porfiri, 2013; Ruberto et al., 2016; Spinello et al., 2019). A two-compartment tank is also

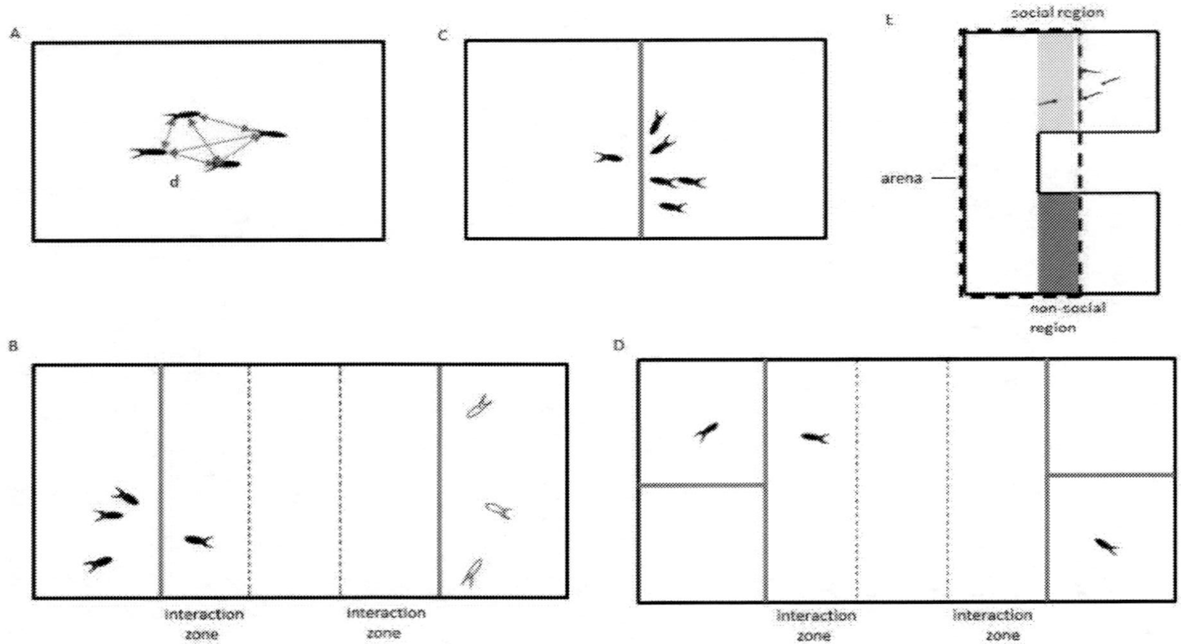

Figure 1. Experiment setup for social interaction assays. (A) Shoaling assay. A group of fish is placed in a tank and their social interactions are tracked. Social cohesion is quantified by measuring the average distance among the members in the shoal. (B) A three-chamber assay to determine the preference of a test fish over two different social stimuli, or preference for a social stimulus vs. non-social cues. (C) A two-chamber assay assessing social preference based on the distance between a test fish and conspecifics. (D) A five-chamber assay for assessing social novelty. (E) A U-shaped three-chamber assay for testing social preference in larvae or young fishes.

employed in some studies (Schroeder et al., 2014), where the proximity between the test fish with the conspecifics is used as a measure of social preference (Figure 1C). Recently, a five-compartment design was implemented in some studies to assess preference to social novelty, by measuring the time the fish spent near 'old' strangers compared to 'new' strangers (Gutiérrez et al., 2019; Vecchia et al., 2019)(Figure 1D). Although less popular, social preference was also assayed in larval and juvenile zebrafish, using a U-shaped three-compartment chamber (Dreosti et al., 2015; Landin et al., 2020) (Figure 1E). A main advantage of social preference test compared to shoaling is the relative ease in quantifying the degree of social preference as fishes are examined individually as opposed to in groups. However, the inability to mingle with conspecifics also posed a restriction to examine how fishes interact naturally.

It has been demonstrated that knockout of *disc1*, an orthologue of human schizophrenia risk gene in zebrafish induced impaired shoaling response to stress (Eachus et al., 2017). Besides, the social behaviors of zebrafish were also found to be susceptible to drugs such as MK-801, lysergic acid diethylamide (LSD), and ketamine (reviewed in Gawel et al., 2019 and Ogi et al., 2021). Such sensitiveness to genetic and pharmacological manipulations suggest that shoaling and social preference could represent ways to model alterations of social interactions that may be relevant to schizophrenia.

Social Aggression

Schizophrenia and aggression are often seen as intimately linked. While aggression has been extensively studied clinically, very little is known about its pathobiology. Adult zebrafish, due to their overt aggressive behaviors, have been used as a model for studying aggression (reviewed in Zabegalov et al., 2019a). A well-established paradigm to provoke aggressive behaviors in the zebrafish is by introducing them a moving attack target, which could be the mirror image of the fish themselves, a dummy fish, or a fish from a video playback (Way et al., 2015). Notably, one putative model of schizophrenia, the ketamine model, has been shown to demonstrate altered aggressive behaviors in the mirror test at a dose-dependent manner (Michelotti et al., 2018).

Another common way to assay aggression is the dyadic fight test – zebrafish dyads are left for a fixed amount of time in a tank and their aggressiveness in the fighting behavior is assessed (Chou et al., 2016; Teles et al., 2016). Generally, the process is monitored by a human observer to keep track of the fight outcomes. Recently, however, Laan, Iglesias-Julios, and de Polavieja (2018) have developed a machine learning algorithm to automatically track and annotate the stereotypical fighting behaviors of individual fish at sub-second resolution, potentially allowing a more nuanced analysis of aggressive behaviors.

Compared to the earlier assay, dyadic fight test is deemed more natural. Remarkably, converging evidence also pointed to a noticeable difference in the hormonal and genomic responses of fishes interacting with real conspecifics compared to their mirror images (Teles et al., 2013; Balzarini et al., 2014; Oliveira et al., 2016), suggesting a need to validate the applicability of mirror test in zebrafish.

Locomotor Hyperactivity

Hyperactivity in animal models is a behavioral measure that has been associated with psychotic agitation (Powell and Miyakawa, 2006), a common manifestation in schizophrenia and other psychiatric diseases. It is believed to arise from a hyperdopaminergic state based on evidence from pharmacology (Gerber et al., 2001) and imaging (Drevets et al., 2001) studies. In animals, hyperactivity is generally assumed from an increase in locomotor activity (Marona-Lewicka, Nichols, and Nichols, 2011; Wierońska et al., 2011).

Depending on the study design, locomotor activity testing in zebrafish has been performed using adults or larvae as young as 4 dpf (Ingebretson and Masino, 2013). Historically, characterization of the activity parameters was scored based on human observation. However, the wide availability of tracking software (reviewed in Franco-Restrepo, Forero, and Vargas, 2019) in recent years has yielded a more comprehensive and reliable analysis of this complex behavioral trait. Multiple parameters have been used to quantify the locomotion of fish, including distance travelled, swimming speed, number of crossings (da Costa Chaulet et al., 2019), number and speed of swim bouts and standstills (Ek et al., 2016), and frequency and duration of swimming episodes (Ingebretson and Masino, 2013).

Numerous reports have been documented for the modulation of locomotor activity by psychomimetic drugs in zebrafish and in rodents, in particular DA agonist (apormorphine, amphetamine) and NMDA receptor antagonists (MK-801, ketamine, and phencyclidine) (Sturgeon, Fessler, and Meltzer, 1979;

Hunt, Raynaud, and Garcia, 2006; Powell and Miyakawa, 2006; Kyzar et al., 2013). It should be noted, however, that the ~~MK801 and~~ MK-801- and ketamine-induced changes occurred in a context- and stress-dependent manner, with hyperactivity observed only when fish were placed in a novel tank (Tran et al., 2016), or when tested in the presence of a mirror image (Michelotti et al., 2018) or a social stimulus (Benvenutti et al., 2021a).

Repetitive Behaviors

Stereotypy is a clinical hallmark of schizophrenia which is often expressed as repetitive speech and other repetitive motor manoeuvres. Although less extensively studied compared to other behavioral phenotypes, several lines of evidence demonstrated that behavior stereotypy in animals can be potentially used as a measure for psychosis-like behaviors (Howes, McCutcheon, and Stone, 2015). In zebrafish, repetitive behaviors are found to manifest naturally, and could be elicited or amplified by genetic or pharmacological manipulations (reviewed in Zabegalov et al., 2019b and Langova et al., 2020; Paredes-Zúñiga, Ormeño, and Allende, 2021).

One distinct repetitive phenotype exhibited by zebrafish is circling, which has been observed in multiple NMDA models of schizophrenia including MK-801, ketamine, and phencyclidine (reviewed in Gawel et al., 2019). Interestingly, a recent study showed that zebrafish also initiated repetitive circling in the absence of illumination (Horstick, Bayleyen, and Burgess, 2020). More recently, Dahlén et al. (2021) showed that the circular swimming elicited by tetrahydrocannabinol was reduced considerably by sulpiride, an antipsychotic for treating schizophrenia, further supporting the notion of repetitive swimming as an indicator of a schizophrenia-like state.

As most of the study on repetitive behaviors in zebrafish were based on manual observations, its reliability warrants a standard metric or an automated analysis method with standardized behavioral endpoints.

Thigmotaxis

Thigmotaxis, also known as 'wall-hugging' or 'wall-following' behavior is the propensity to move or stay near the edge as opposed to the centre of a novel environment. This behavior has been commonly observed in a wide range of species including rodents (Prut and Belzung, 2003), fish (Schnörr et al., 2012), and humans (Kallai et al., 2007; Walz, Mühlberger, and Pauli, 2016) in the laboratory settings, and is thought to be an anxiety-related behavioral response. When placed in a square or a circular tank, adult zebrafish consistently showed a strong tendency to avoid the centre arena and swim near the wall of the tank (Champagne et al., 2010; Buske and Gerlai, 2012). Experiments with larval zebrafish have nevertheless yielded mixed results. As opposed to studies which showed an increased percentage of distance moved in the outer zone of multi-well plates when light was off (Schnörr et al., 2012; Best and Vijayan, 2018), Horstick et al. (2017) observed opposite results from larvae housed in a concave arena. It was demonstrated that the larvae which originally swam near the wall avoided themselves from edges after the loss of illumination, suggesting that the behavior could represent a tendency to swim outwards in relation to luminance changes instead of preference for edges.

Increased thigmotaxis has been induced with exposure to several pharmacological drugs such as nicotine, and in zebrafish with *dat* knockout (reviewed in Zabegalov et al., 2019b). Although potentially suggests an elevation in the anxiety level, the thigmotaxic response in zebrafish may represent a pathological repetitive behavior, similar to stereotypic peripheral hyperlocomotion seen in rodents (Hodgson et al., 2010; Zabegalov et al., 2019b). More comprehensive experiments using anxiolytic drugs remain to be performed to validate the anxiety-like nature of thigmotaxic behavior in zebrafish.

Bottom Dwelling

The initial tendency to dive to the bottom part of a novel experimental tank followed by a gradual increase in upwards vertical activity is a behavior of zebrafish known as bottom dwelling. Together with thigmotaxis, this inhibited exploration behavior is regarded as the attempts of fish to 'cling' to a solid object to avoid predators (Gerlai, 2010). Bottom dwelling behavior in zebrafish is typically assayed using a simple novel tank diving test (Levin, Bencan, and Cerutti, 2007), where the preferred position of the fish in the tank and the latency to explore the upper part of the tank are tracked and scored as an index of anxiety. A cylindrical tank is sometimes used as an alternative to a simple tank to ease the monitoring of the swimming behaviors (Ek et al., 2016; Kalueff, 2017).

The anxiety-related nature of bottom dwelling behavior in zebrafish has been validated through pharmacological manipulation with multiple anxiolytic drugs (Sackerman et al., 2010; Hawkey et al., 2021). More recently, zebrafish with homozygous loss of DAT (Wang et al., 2019) and with disrupted *disc1* gene (García-González et al., 2021) – two mutations that have been implicated to associate with schizophrenia – were shown to express an increased bottom 'digging' or dwelling response, which may indicate an increased anxiety-like behavior. Overall, these findings suggest that the bottom dwelling behavior is highly responsive to pharmacological and genetic disruptions, indicating the utility of the novel tank diving test for unveiling anxiety circuitry in zebrafish schizophrenia models.

Zebrafish Modelling of Schizophrenia: Current and Future Directions

The use of zebrafish models for the study of schizophrenia has begun in earnest in recent years. However, most of these studies involved modelling of single risk gene at a particular developmental stage, and investigating relatively limited anatomical, functional, and behavioral phenotypes, greatly restricting the translational power of the models. To advance our understanding of schizophrenia biology, it will be important to go on more concerted and thorough approaches to delineate the causal links among genes, neural circuits, and behavioral outputs. These endeavours, although many still in their infancy, promise additional insights into the pathophysiology of this complex disease. This section highlights a few key directions of these endeavours.

Large-Scale Mutant Phenotyping

Mutant phenotyping is a pivotal step in identifying the key genes underlying disease etiology and their neurobiological functions. Recent advancement in the field of genetics and genomics has overcome many hurdles and limitations in generating targeted mutations in genes of interest, enabling manipulation of zebrafish genome with remarkable ease and efficiency (Zhuo et al., 2017; Cano-Gamez and Trynka, 2020). Recently, Thyme et al. (2019) have generated and analyzed 132 mutants for genes found in the schizophrenia-associated genomic regions identified by GWAS. By performing high-throughput assays to systematically examine and compare the brain structure, functional activity, and behaviors of these mutants, they have gathered a rich set of phenotypic data that was powerful enough to identify high-confidence risk genes in the implicated loci. Remarkably, recent development in high-throughput behavioral assays (Joo et al., 2021) and machine learning algorithms for behavioral tracking (Sun et al., 2019; Yang et al., 2021) are also promising to expedite the progress in large-scale mutant phenotyping. Future developments should focus on further improving the sensitivity and ease of use of these assays for drug screening, and ultimately for development of precision medicine for schizophrenia treatment.

Functional Imaging of Neuronal Activity

Functional imaging is a powerful technique that enables the studies of brain dynamics in zebrafish coupled with genetic fluorescence reporters. To capture the whole-brain activity, zebrafish are normally embedded in agarose, either fully or partially to hold their heads still. Imaging was performed using either confocal or two-photon microscopes (Portugues et al., 2014; Takahashi et al., 2017; Light and Jontes, 2019), and more recently also with light-sheet microscopes (Liu et al., 2019) to yield the brain activity at single cell resolution. Notably, Takahashi, Narushima, and Oda (2002), Kohashi and Oda (2008), and Takahashi et al. (2017) have successfully demonstrated the neural basis underlying escape behavior and habituation by imaging the activity of Mauthner and non-Mauthner cells in larval zebrafish. Although these preparations were often employed on young larvae up to 8 dpf, several recent studies have shown that robust functional activity can also be detected in the late-stage larvae in the brain regions governing olfaction, gustation, and vision (Jetti, Vendrell-Llopis, and Yaksi, 2014; Vendrell-Llopis and Yaksi, 2015; Matsuda et al., 2017; Bergmann et al., 2018). Since many forms of sensory-driven behaviors in zebrafish emerged at later developmental stages, these techniques open an avenue for revealing the functional maps that link neuronal activities and behaviors.

Activity Mapping Using Neural-Activity Markers

Although it is possible to generate brain-wide activity maps using functional imaging, the behavioral activity that could be studied – using either the full or the partially restrained fish techniques – are limited. Neural activity markers such as immediate early gene *c-fos* (Baraban et al., 2005) and phosphorylated extracellular-signaling-regulated kinase (pERK)(Randlett et al., 2015) present as useful tools for anatomical functional mapping in freely swimming zebrafish, potentially allowing studies of nearly all behaviors. These mapping methods, although not real time and not of high resolution, accurately reveal the activation regions associated with behaviors and stimuli. These high-throughput techniques have rendered whole-brain activity map in larval zebrafish carrying mutations in hundreds of schizophrenia-associated genes (Thyme et al., 2019). The use of activity markers has also allowed researchers to successfully outline the brain regions associated with social preference (Tunbak et al., 2020), and offer potential to identify cells and circuits driving behaviors relevant to other neurological diseases.

Perturbation of Neural Circuitry

Perturbation of neuronal circuits plays a key role in neuroscience research, providing additional insights that maybe otherwise difficult to decipher from anatomical and functional studies. Perturbation of neuronal activity in zebrafish has been successfully achieved with cell ablation and optogenetic stimulation or silencing, which involves either single neuron or a group of neurons (Haesemeyer and Schier, 2015). In a recent study, Hecker et al. (2020) showed that by ablating the giant Mauthner cell from zebrafish, including its axon, the rapid escape of the fish was also lost irreversibly, suggesting the essential role of Mauthner cells in mediating escape behavior that was long thought to be well-driven by other small neurons. These approaches hold promise in addressing the effects of brain lesions on the clinical deficits manifested in neurological diseases, advancing our understandings of the functional roles of neuron and its circuits.

Imaging of Freely Behaving Zebrafish

Visualization of neural activity in freely behaving animals is a crucial step towards understanding how brains generate behaviors. Recently, Muto and Kawakami (2016) showed that it is possible to image neural activity during prey capture behaviors in unrestrained larvae using a novel method developed around a simple epifluorescence microscope. The recordings were, however, limited to the still periods between the swimming bouts. Techniques for continuous imaging of neural activity in young larvae were later introduced by Cong et al. (2017) and Kim et al. (2017), by integrating imaging and sophisticated tracking methods to keep the target fish in focus. These methods have greatly expanded the spectrum of natural behaviors amenable to functional imaging, and could potentially shed light on the behavioral circuits relevant to schizophrenia.

Electrophysiology

The existing electrophysiological tools for studying neuronal activity of aquatic animals have been limited, especially for small-size animals like zebrafish. Several research groups, however, have successfully recorded the brain activity from larval (Zdebik et al., 2013; Meyer et al., 2016; Tomasello and Sive, 2020) and adult (Pineda, Beattie, and Hall, 2011; Cho et al., 2017) zebrafish, using electrophysiological techniques which are either invasive or non-invasive in nature. These setups have enabled long-term recordings of electrophysiological activity up to hours in individual fish. Simultaneous recordings of local field potentials from multiple zebrafish have also recently been made possible, through an electrophysiological platform integrated with microfluidic technology (Hong et al., 2016). These techniques have been shown to be promising for electrophysiological characterization of epileptic zebrafish and screening of antiepileptic drugs, and are likely to have broad applications to other neurological disease models. The high temporal resolution of electrophysiological signals could potentially offer additional insights into neural dynamics, complementing the details that could not be obtained using functional imaging.

Conclusion

Dissecting the genetic basis and the circuitry underlying complex mental illness is a challenging endeavor. As the achievement in the previous work suggests, zebrafish is rapidly emerging as a powerful tool for modelling genetic risk and impairment in fundamental neural pathways and basic behavioral circuits associated with schizophrenia. Together with the development of novel assays and improved procedures on cross-species validation, the application of this model is likely to yield critical insights translatable to human studies.

Acknowledgments

The work in authors' lab was supported by grants from the Fundamental Research Grant Scheme under the Ministry of Higher Education, Malaysia (FRGS/1/2018/STG03/UTAR/02/1) and the Malaysia Toray Science Foundation (MTSF grant 4417/M01).

References

Abi-Dargham, A. (2007). Alterations of serotonin transmission in schizophrenia. *International review of neurobiology*, *78*, 133-164.

Akbarian, S., Kim, J. J., Potkin, S. G., Hagman, J. O., Tafazzoli, A., Bunney, W. E., and Jones, E. G. (1995). Gene expression for glutamic acid decarboxylase is reduced without loss of neurons in prefrontal cortex of schizophrenics. *Archives of general psychiatry*, *52*(4), 258-266.

Amador, X. F. and Gorman, J. M. (1998). Psychopathologic domains and insight in schizophrenia. *Psychiatric Clinics of North America*, *21*(1), 27-42.

Amsterdam, A., Varshney, G. K., and Burgess, S. M. (2011). Retroviral-mediated insertional mutagenesis in zebrafish. *Methods in cell biology*, *104*, 59-82.

Andersson, M. Å., Ek, F., and Olsson, R. (2015). Using visual lateralization to model learning and memory in zebrafish larvae. *Scientific reports*, *5*(1), 1-7.

Arneric, S. P., Sullivan, J. P., & Williams, M. (1995). Neuronal nicotinic acetylcholine receptors: Novel targets for central nervous system therapeutics. *Psychopharmacology: the fourth generation of progress*, 95-110.

Asada, H., Kawamura, Y., Maruyama, K., Kume, H., Ding, R. G., Ji, F. Y., ... and Obata, K. (1996). Mice lacking the 65 kDa isoform of glutamic acid decarboxylase (GAD65) maintain normal levels of GAD67 and GABA in their brains but are susceptible to seizures. *Biochemical and biophysical research communications*, *229*(3), 891-895.

Bak, L. K., Schousboe, A., and Waagepetersen, H. S. (2006). The glutamate/GABA-glutamine cycle: aspects of transport, neurotransmitter homeostasis and ammonia transfer. *Journal of neurochemistry*, *98*(3), 641-653.

Balzarini, V., Taborsky, M., Wanner, S., Koch, F., and Frommen, J. G. (2014). Mirror, mirror on the wall: the predictive value of mirror tests for measuring aggression in fish. *Behavioral Ecology and Sociobiology*, *68*(5), 871-878.

Banono, N. S. and Esguerra, C. V. (2020). Pharmacological validation of the prepulse inhibition of startle response in larval zebrafish using a commercial automated system and software. *JoVE (Journal of Visualized Experiments)*, (161), e61423.

Banono, N. S., Gawel, K., De Witte, L., and Esguerra, C. V. (2021). Zebrafish larvae carrying a splice variant mutation in cacna1d: a new model for schizophrenia-like behaviours?. *Molecular neurobiology*, *58*(2), 877-894.

Baraban, S. C., Taylor, M. R., Castro, P. A., and Baier, H. (2005). Pentylenetetrazole induced changes in zebrafish behavior, neural activity and c-fos expression. *Neuroscience*, *131*(3), 759-768.

Barnes, N. M. and Sharp, T. (1999). A review of central 5-HT receptors and their function. *Neuropharmacology*, *38*(8), 1083-1152.

Bartolini, T., Mwaffo, V., Showler, A., Macrì, S., Butail, S., and Porfiri, M. (2016). Zebrafish response to 3D printed shoals of conspecifics: the effect of body size. *Bioinspiration and biomimetics*, *11*(2), 026003.

Barut, B. A. and Zon, L. I. (2000). Realizing the potential of zebrafish as a model for human disease. *Physiological Genomics*, *2*(2), 49-51.

Bayle, F. J., Leroy, S., Gourion, D., Millet, B., Olie, J. P., Poirier, M. F., and Krebs, M. O. (2003). 5HTTLPR polymorphism in schizophrenic patients: further support for association with violent suicide attempts. *American Journal of Medical Genetics Part B: Neuropsychiatric Genetics*, *119*(1), 13-17.

Beaulieu, J. M. and Gainetdinov, R. R. (2011). The physiology, signaling, and pharmacology of dopamine receptors. *Pharmacological reviews*, *63*(1), 182-217.

Benes, F. M. (2009). Neural circuitry models of schizophrenia: is it dopamine, GABA, glutamate, or something else?. *Biological psychiatry*, *65*(12), 1003-1005.

Benvenutti, R., Gallas-Lopes, M., Sachett, A., Marcon, M., Strogulski, N. R., Reis, C. G., ... and Herrmann, A. P. (2021a). How do zebrafish (Danio rerio) respond to MK-801 and amphetamine? Relevance for assessing schizophrenia-related endophenotypes in alternative model organisms. *Journal of Neuroscience Research*.

Benvenutti, R., Marcon, M., Gallas-Lopes, M., de Melo, A. J., Herrmann, A. P., and Piato, A. (2021b). Swimming in the maze: An overview of maze apparatuses and protocols to assess zebrafish behavior. *Neuroscience and Biobehavioral Reviews*.

Bergeron, S. A., Carrier, N., Li, G. H., Ahn, S., and Burgess, H. A. (2015). Gsx1 expression defines neurons required for prepulse inhibition. *Molecular psychiatry*, *20*(8), 974-985.

Bergmann, K., Meza Santoscoy, P., Lygdas, K., Nikolaeva, Y., MacDonald, R. B., Cunliffe, V. T., and Nikolaev, A. (2018). Imaging neuronal activity in the optic tectum of late stage larval zebrafish. *Journal of developmental biology*, *6*(1), 6.

Best, C. and Vijayan, M. M. (2018). Cortisol elevation post-hatch affects behavioural performance in zebrafish larvae. *General and comparative endocrinology*, *257*, 220-226.

Bettler, B., Kaupmann, K., Mosbacher, J., and Gassmann, M. (2004). Molecular structure and physiological functions of GABAB receptors. *Physiological reviews*, *84*(3), 835-867.

Bhandiwad, A. A., Zeddies, D. G., Raible, D. W., Rubel, E. W., and Sisneros, J. A. (2013). Auditory sensitivity of larval zebrafish (Danio rerio) measured using a behavioral prepulse inhibition assay. *Journal of Experimental Biology*, *216*(18), 3504-3513.

Bhugra, D. (2005). The global prevalence of schizophrenia. *PLoS medicine*, *2*(5), e151.

Bilic, P., Jukic, V., Vilibic, M., Savic, A., and Bozina, N. (2014). Treatment-resistant schizophrenia and DAT and SERT polymorphisms. *Gene*, *543*(1), 125-132.

Bill, B. R., Petzold, A. M., Clark, K. J., Schimmenti, L. A., and Ekker, S. C. (2009). A primer for morpholino use in zebrafish. *Zebrafish*, *6*(1), 69-77.

Blanchard, J. J., Kring, A. M., Horan, W. P., and Gur, R. (2011). Toward the next generation of negative symptom assessments: the collaboration to advance negative symptom assessment in schizophrenia. *Schizophrenia bulletin*, *37*(2), 291-299.

Blank, M., Guerim, L. D., Cordeiro, R. F., and Vianna, M. R. (2009). A one-trial inhibitory avoidance task to zebrafish: rapid acquisition of an NMDA-dependent long-term memory. *Neurobiology of learning and memory*, *92*(4), 529-534.

Blechman, J., Borodovsky, N., Eisenberg, M., Nabel-Rosen, H., Grimm, J., and Levkowitz, G. (2007). Specification of hypothalamic neurons by dual regulation of the homeodomain protein Orthopedia.

Boehmler, W., Obrecht-Pflumio, S., Canfield, V., Thisse, C., Thisse, B., and Levenson, R. (2004). Evolution and expression of D2 and D3 dopamine receptor genes in zebrafish. *Developmental dynamics: an official publication of the American Association of Anatomists*, *230*(3), 481-493.

Boehmler, W., Carr, T., Thisse, C., Thisse, B., Canfield, V. A., and Levenson, R. (2007). D4 Dopamine receptor genes of zebrafish and effects of the antipsychotic clozapine on larval swimming behaviour. *Genes, Brain and Behavior*, *6*(2), 155-166.

Braff, D. L., Geyer, M. A., and Swerdlow, N. R. (2001). Human studies of prepulse inhibition of startle: normal subjects, patient groups, and pharmacological studies. *Psychopharmacology*, *156*(2), 234-258.

Brandon, N. J. and Sawa, A. (2011). Linking neurodevelopmental and synaptic theories of mental illness through DISC1. *Nature Reviews Neuroscience*, *12*(12), 707-722.

Breier, A. (1995). Serotonin, schizophrenia and antipsychotic drug action. *Schizophrenia research*, *14*(3), 187-202.

Bullock, W. M., Cardon, K., Bustillo, J., Roberts, R. C., and Perrone-Bizzozero, N. I. (2008). Altered expression of genes involved in GABAergic transmission and neuromodulation of granule cell activity in the cerebellum of schizophrenia patients. *American Journal of Psychiatry*, *165*(12), 1594-1603.

Burgess, H. A. and Granato, M. (2007). Sensorimotor gating in larval zebrafish. *Journal of Neuroscience*, *27*(18), 4984-4994.

Buske, C. and Gerlai, R. (2012). Maturation of shoaling behavior is accompanied by changes in the dopaminergic and serotoninergic systems in zebrafish. *Developmental psychobiology*, *54*(1), 28-35.

Cano-Gamez, E. and Trynka, G. (2020). From GWAS to function: using functional genomics to identify the mechanisms underlying complex diseases. *Frontiers in Genetics*, *11*, 424.

Caspi, A. and Moffitt, T. E. (2006). Gene–environment interactions in psychiatry: joining forces with neuroscience. *Nature Reviews Neuroscience*, *7*(7), 583-590.

Caulfield, M. P.,and Birdsall, N. J. (1998). International Union of Pharmacology. XVII. Classification of muscarinic acetylcholine receptors. *Pharmacological reviews*, *50*(2), 279-290.

Champagne, D. L., Hoefnagels, C. C., De Kloet, R. E., and Richardson, M. K. (2010). Translating rodent behavioral repertoire to zebrafish (Danio rerio): relevance for stress research. *Behavioural brain research, 214*(2), 332-342.

Charnay, Y. and Léger, L. (2010). Brain serotonergic circuitries. *Dialogues in clinical neuroscience, 12*(4), 471.

Chen, L., Stuart, L., Ohsumi, T. K., Burgess, S., Varshney, G. K., Dastur, A., ... and Schmidt, E. V. (2013). Transposon activation mutagenesis as a screening tool for identifying resistance to cancer therapeutics. *BMC cancer, 13*(1), 1-15.

Cheng, Y. C., Scotting, P. J., Hsu, L. S., Lin, S. J., Shih, H. Y., Hsieh, F. Y., ... and Shen, C. J. (2013). Zebrafish rgs4 is essential for motility and axonogenesis mediated by Akt signaling. *Cellular and Molecular Life Sciences, 70*(5), 935-950.

Chiang, C., Litingtung, Y., Lee, E., Young, K. E., Corden, J. L., Westphal, H., and Beachy, P. A. (1996). Cyclopia and defective axial patterning in mice lacking Sonic hedgehog gene function. *Nature, 383*(6599), 407-413.

Cho, S. J., Byun, D., Nam, T. S., Choi, S. Y., Lee, B. G., Kim, M. K., and Kim, S. (2017). Zebrafish as an animal model in epilepsy studies with multichannel EEG recordings. *Scientific reports, 7*(1), 1-10.

Chou, M. Y., Amo, R., Kinoshita, M., Cherng, B. W., Shimazaki, H., Agetsuma, M., ... and Okamoto, H. (2016). Social conflict resolution regulated by two dorsal habenular subregions in zebrafish. *Science, 352*(6281), 87-90.

Chubb, J. E., Bradshaw, N. J., Soares, D. C., Porteous, D. J., and Millar, J. K. (2008). The DISC locus in psychiatric illness. *Molecular psychiatry, 13*(1), 36-64.

Ciliax, B. J., Heilman, C., Demchyshyn, L. L., Pristupa, Z. B., Ince, E., Hersch, S. M., ... and Levey, A. I. (1995). The dopamine transporter: immunochemical characterization and localization in brain. *Journal of Neuroscience, 15*(3), 1714-1723.

Ciliax, B. J., Drash, G. W., Staley, J. K., Haber, S., Mobley, C. J., Miller, G. W., ... and Levey, A. I. (1999). Immunocytochemical localization of the dopamine transporter in human brain. *Journal of Comparative Neurology, 409*(1), 38-56.

Cleal, M., Fontana, B. D., Ranson, D. C., McBride, S. D., Swinny, J. D., Redhead, E. S., and Parker, M. O. (2021). The Free-movement pattern Y-maze: A cross-species measure of working memory and executive function. *Behavior Research Methods, 53*(2), 536-557.

Cocco, A., Rönnberg, A. C., Jin, Z., André, G. I., Vossen, L. E., Bhandage, A. K., ... and Winberg, S. (2017). Characterization of the γ-aminobutyric acid signaling system in the zebrafish (Danio rerio Hamilton) central nervous system by reverse transcription-quantitative polymerase chain reaction. *Neuroscience, 343*, 300-321.

Cognato, G. D. P., Bortolotto, J. W., Blazina, A. R., Christoff, R. R., Lara, D. R., Vianna, M. R., and Bonan, C. D. (2012). Y-Maze memory task in zebrafish (Danio rerio): the role of glutamatergic and cholinergic systems on the acquisition and consolidation periods. *Neurobiology of learning and memory, 98*(4), 321-328.

Collins, J. E., White, S., Searle, S. M., and Stemple, D. L. (2012). Incorporating RNA-seq data into the zebrafish Ensembl genebuild. *Genome research, 22*(10), 2067-2078.

Cong, L., Ran, F. A., Cox, D., Lin, S., Barretto, R., Habib, N., ... and Zhang, F. (2013). Multiplex genome engineering using CRISPR/Cas systems. *Science, 339*(6121), 819-823.

Cong, L., Wang, Z., Chai, Y., Hang, W., Shang, C., Yang, W., ... and Wen, Q. (2017). Rapid whole brain imaging of neural activity in freely behaving larval zebrafish (Danio rerio). *Elife*, *6*, e28158.

Connaughton, V. P., Nelson, R., and Bender, A. M. (2008). Electrophysiological evidence of GABAA and GABAC receptors on zebrafish retinal bipolar cells. *Visual neuroscience*, *25*(2), 139-153.

Corley, S. M., Tsai, S. Y., Wilkins, M. R., and Shannon Weickert, C. (2016). Transcriptomic analysis shows decreased cortical expression of NR4A1, NR4A2 and RXRB in schizophrenia and provides evidence for nuclear receptor dysregulation. *PloS one*, *11*(12), e0166944.

Court, J., Spurden, D., Lloyd, S., McKeith, I., Ballard, C., Cairns, N., ... and Perry, E. (1999). Neuronal Nicotinic Receptors in Dementia with Lewy Bodies and Schizophrenia: α-Bungarotoxin and Nicotine Binding in the Thalamus. *Journal of neurochemistry*, *73*(4), 1590-1597.

Court, J. A., Piggott, M. A., Lloyd, S., Cookson, N., Ballard, C. G., McKeith, I. G., ... and Perry, E. K. (2000). Nicotine binding in human striatum: elevation in schizophrenia and reductions in dementia with Lewy bodies, Parkinson's disease and Alzheimer's disease and in relation to neuroleptic medication. *Neuroscience*, *98*(1), 79-87.

Coyle, J. T. (2012). NMDA receptor and schizophrenia: a brief history. *Schizophrenia bulletin*, *38*(5), 920-926.

Cunningham, F., Amode, M. R., Barrell, D., Beal, K., Billis, K., Brent, S., ... and Flicek, P. (2015). Ensembl 2015. *Nucleic acids research*, *43*(D1), D662-D669.

Curley, A. A., Arion, D., Volk, D. W., Asafu-Adjei, J. K., Sampson, A. R., Fish, K. N., and Lewis, D. A. (2011). Cortical deficits of glutamic acid decarboxylase 67 expression in schizophrenia: clinical, protein, and cell type-specific features. *American Journal of Psychiatry*, *168*(9), 921-929.

da Costa Chaulet, F., de Alcantara Barcellos, H. H., Fior, D., Pompermaier, A., Koakoski, G., da Rosa, J. G. S., ... and Barcellos, L. J. G. (2019). Glyphosate-and fipronil-based agrochemicals and their mixtures change zebrafish behavior. *Archives of environmental contamination and toxicology*, *77*(3), 443-451.

Dahlén, A., Zarei, M., Melgoza, A., Wagle, M., and Guo, S. (2021). THC-induced behavioral stereotypy in zebrafish as a model of psychosis-like behavior. *Scientific reports*, *11*(1), 1-13.

Dayan, P. and Huys, Q. J. M. (2008). Serotonin, inhibition, and negative mood. *PLoS computational biology*, *4*(2), e4.

Dean, B., Crook, J. M., Opeskin, K., Hill, C., Keks, N., and Copolov, D. L. (1996). The density of muscarinic M1 receptors is decreased in the caudate-putamen of subjects with schizophrenia. *Molecular psychiatry*, *1*(1), 54-58.

Delgado, L. and Schmachtenberg, O. (2008). Immunohistochemical localization of GABA, GAD65, and the receptor subunits GABA Aα1 and GABA B1 in the zebrafish cerebellum. *The Cerebellum*, *7*(3), 444-450.

Doyon, Y., McCammon, J. M., Miller, J. C., Faraji, F., Ngo, C., Katibah, G. E., ... and Amacher, S. L. (2008). Heritable targeted gene disruption in zebrafish using designed zinc-finger nucleases. *Nature biotechnology*, *26*(6), 702-708.

Drago, A., De Ronchi, D., and Serretti, A. (2008). 5-HT 1A gene variants and psychiatric disorders: a review of current literature and selection of SNPs for future studies. *The The International Journal of Neuropsychopharmacology*, *11*(5), 701-721.

Draper, B. W., McCallum, C. M., Stout, J. L., Slade, A. J., and Moens, C. B. (2004). A high-throughput method for identifying N-ethyl-N-nitrosourea (ENU)-induced point mutations in zebrafish. *Methods in cell biology*, *77*, 91-112.

Dreosti, E., Lopes, G., Kampff, A. R., and Wilson, S. W. (2015). Development of social behavior in young zebrafish. *Frontiers in neural circuits*, *9*, 39.

Drevets, W. C., Gautier, C., Price, J. C., Kupfer, D. J., Kinahan, P. E., Grace, A. A., ... and Mathis, C. A. (2001). Amphetamine-induced dopamine release in human ventral striatum correlates with euphoria. *Biological psychiatry*, *49*(2), 81-96.

Eachus, H., Bright, C., Cunliffe, V. T., Placzek, M., Wood, J. D., and Watt, P. J. (2017). Disrupted-in-Schizophrenia-1 is essential for normal hypothalamic-pituitary-interrenal (HPI) axis function. *Human molecular genetics*, *26*(11), 1992-2005.

Eglen, R. M. and Watson, N. (1996). Selective muscarinic receptor agonists and antagonists. *Pharmacology and toxicology*, *78*(2), 59-68.

Eguiraun, H., Casquero, O., Sørensen, A. J., and Martinez, I. (2018). Reducing the Number of Individuals to Monitor Shoaling Fish Systems–Application of the Shannon Entropy to Construct a Biological Warning System Model. *Frontiers in physiology*, *9*, 493.

Eisen, J. S. and Smith, J. C. (2008). Controlling morpholino experiments: don't stop making antisense.

Ek, F., Malo, M., Åberg Andersson, M., Wedding, C., Kronborg, J., Svensson, P., ... and Olsson, R. (2016). Behavioral analysis of dopaminergic activation in zebrafish and rats reveals similar phenotypes. *ACS chemical neuroscience*, *7*(5), 633-646.

Emamian, E. S., Hall, D., Birnbaum, M. J., Karayiorgou, M., and Gogos, J. A. (2004). Convergent evidence for impaired AKT1-GSK3β signaling in schizophrenia. *Nature genetics*, *36*(2), 131-137.

Esteves, P. J., Abrantes, J., Baldauf, H. M., BenMohamed, L., Chen, Y., Christensen, N., ... and Mage, R. (2018). The wide utility of rabbits as models of human diseases. *Experimental and molecular medicine*, *50*(5), 1-10.

Everitt, B. J. and Robbins, T. W. (1997). Central cholinergic systems and cognition. *Annual review of psychology*, *48*(1), 649-684.

Facciol, A. and Gerlai, R. (2020). Zebrafish Shoaling, Its Behavioral and Neurobiological Mechanisms, and Its Alteration by Embryonic Alcohol Exposure: A Review. *Frontiers in Behavioral Neuroscience*, *14*, 174.

Fiori, L. M. and Turecki, G. (2008). Implication of the polyamine system in mental disorders. *Journal of psychiatry and neuroscience: JPN*, *33*(2), 102.

Force, A., Lynch, M., Pickett, F. B., Amores, A., Yan, Y. L., and Postlethwait, J. (1999). Preservation of duplicate genes by complementary, degenerative mutations. *Genetics*, *151*(4), 1531-1545.

Formella, I., Scott, E. K., Burne, T. H., Harms, L. R., Liu, P. Y., Turner, K. M., ... and Eyles, D. W. (2012). Transient knockdown of tyrosine hydroxylase during development has persistent effects on behaviour in adult zebrafish (Danio rerio).

Franco-Restrepo, J. E., Forero, D. A., and Vargas, R. A. (2019). A review of freely available, open-source software for the automated analysis of the behavior of adult zebrafish. *Zebrafish, 16*(3), 223-232.

Freedman, R., Hall, M., Adler, L. E., and Leonard, S. (1995). Evidence in postmortem brain tissue for decreased numbers of hippocampal nicotinic receptors in schizophrenia. *Biological psychiatry, 38*(1), 22-33.

Fremeau, R. T., Burman, J., Qureshi, T., Tran, C. H., Proctor, J., Johnson, J., ... and Edwards, R. H. (2002). The identification of vesicular glutamate transporter 3 suggests novel modes of signaling by glutamate. *Proceedings of the National Academy of Sciences, 99*(22), 14488-14493.

Fremeau Jr, R. T., Voglmaier, S., Seal, R. P., and Edwards, R. H. (2004). VGLUTs define subsets of excitatory neurons and suggest novel roles for glutamate. *Trends in neurosciences, 27*(2), 98-103.

Fromer, M., Roussos, P., Sieberts, S. K., Johnson, J. S., Kavanagh, D. H., Perumal, T. M., ... and Sklar, P. (2016). Gene expression elucidates functional impact of polygenic risk for schizophrenia. *Nature neuroscience, 19*(11), 1442-1453.

Fu, Y., Foden, J. A., Khayter, C., Maeder, M. L., Reyon, D., Joung, J. K., and Sander, J. D. (2013). High-frequency off-target mutagenesis induced by CRISPR-Cas nucleases in human cells. *Nature biotechnology, 31*(9), 822-826.

Furness, D. N., Dehnes, Y., Akhtar, A. Q., Rossi, D. J., Hamann, M., Grutle, N. J., ... and Danbolt, N. C. (2008). A quantitative assessment of glutamate uptake into hippocampal synaptic terminals and astrocytes: new insights into a neuronal role for excitatory amino acid transporter 2 (EAAT2). *Neuroscience, 157*(1), 80-94.

García-González, J., Quadros, B. D., Havelange, W., Brock, A. J., and Brennan, C. H. (2021). Behavioral Effects of Developmental Exposure to JWH-018 in Wild-Type and Disrupted in Schizophrenia 1 (disc1) Mutant Zebrafish. *Biomolecules, 11*(2), 319.

Gauthier, J., Champagne, N., Lafrenière, R. G., Xiong, L., Spiegelman, D., Brustein, E., ... and S2D Team. (2010). De novo mutations in the gene encoding the synaptic scaffolding protein SHANK3 in patients ascertained for schizophrenia. *Proceedings of the National Academy of Sciences, 107*(17), 7863-7868.

Gawel, K., Banono, N. S., Michalak, A., and Esguerra, C. V. (2019). A critical review of zebrafish schizophrenia models: Time for validation?. *Neuroscience and Biobehavioral Reviews, 107*, 6-22.

Gerber, D. J., Sotnikova, T. D., Gainetdinov, R. R., Huang, S. Y., Caron, M. G., and Tonegawa, S. (2001). Hyperactivity, elevated dopaminergic transmission, and response to amphetamine in M1 muscarinic acetylcholine receptor-deficient mice. *Proceedings of the National Academy of Sciences, 98*(26), 15312-15317.

Gerlai, R. (2010). Zebrafish antipredatory responses: a future for translational research?. *Behavioural brain research, 207*(2), 223-231.

Gerlai, R. (2016). Learning and memory in zebrafish (Danio rerio). *Methods in Cell Biology, 134*, 551-586.

Gerlai, R. (2017). Animated images in the analysis of zebrafish behavior. *Current zoology, 63*(1), 35-44.

Geyer, M. A., McIlwain, K. L., and Paylor, R. (2002). Mouse genetic models for prepulse inhibition: an early review. *Molecular psychiatry, 7*(10), 1039-1053.

Giacomotto, J., Carroll, A. P., Rinkwitz, S., Mowry, B., Cairns, M. J., and Becker, T. S. (2016). Developmental suppression of schizophrenia-associated miR-137 alters sensorimotor function in zebrafish. *Translational psychiatry*, *6*(5), e818-e818.

Golimbet, V. E., Alfimova, M. V., Shchebatykh, T. V., Abramova, L. I., Kaleda, V. G., and Rogaev, E. I. (2004). Serotonin transporter polymorphism and depressive-related symptoms in schizophrenia. *American Journal of Medical Genetics Part B: Neuropsychiatric Genetics*, *126*(1), 1-7.

Golzio, C., Willer, J., Talkowski, M. E., Oh, E. C., Taniguchi, Y., Jacquemont, S., ... and Katsanis, N. (2012). KCTD13 is a major driver of mirrored neuroanatomical phenotypes of the 16p11. 2 copy number variant. *Nature*, *485*(7398), 363-367.

Gotti, C., Clementi, F., Fornari, A., Gaimarri, A., Guiducci, S., Manfredi, I., ... and Zoli, M. (2009). Structural and functional diversity of native brain neuronal nicotinic receptors. *Biochemical pharmacology*, *78*(7), 703-711.

Gras, C., Herzog, E., Bellenchi, G. C., Bernard, V., Ravassard, P., Pohl, M., ... and El Mestikawy, S. (2002). A third vesicular glutamate transporter expressed by cholinergic and serotoninergic neurons. *Journal of Neuroscience*, *22*(13), 5442-5451.

Guan, Z. Z., Zhang, X., Ravid, R., and Nordberg, A. (2000). Decreased protein levels of nicotinic receptor subunits in the hippocampus and temporal cortex of patients with Alzheimer's disease. *Journal of neurochemistry*, *74*(1), 237-243.

Gurevich, V. V. and Gurevich, E. V. (2006). The structural basis of arrestin-mediated regulation of G-protein-coupled receptors. *Pharmacology and therapeutics*, *110*(3), 465-502.

Gusev, A., Mancuso, N., Won, H., Kousi, M., Finucane, H. K., Reshef, Y., ... and Price, A. L. (2018). Transcriptome-wide association study of schizophrenia and chromatin activity yields mechanistic disease insights. *Nature genetics*, *50*(4), 538-548.

Gutiérrez, H. C., Colanesi, S., Cooper, B., Reichmann, F., Young, A. M., Kelsh, R. N., and Norton, W. H. (2019). Endothelin neurotransmitter signalling controls zebrafish social behaviour. *Scientific reports*, *9*(1), 1-17.

Haesemeyer, M. and Schier, A. F. (2015). The study of psychiatric disease genes and drugs in zebrafish. *Current opinion in neurobiology*, *30*, 122-130.

Haffter, P., Granato, M., Brand, M., Mullins, M. C., Hammerschmidt, M., Kane, D. A., ... and Nusslein-Volhard, C. (1996). The identification of genes with unique and essential functions in the development of the zebrafish, Danio rerio. *Development*, *123*(1), 1-36.

Han, S. O., Kommaddi, R. P., and Shenoy, S. K. (2013). Distinct roles for β-arrestin2 and arrestin-domain-containing proteins in β2 adrenergic receptor trafficking. *EMBO reports*, *14*(2), 164-171.

Hashimoto, T., Volk, D. W., Eggan, S. M., Mirnics, K., Pierri, J. N., Sun, Z., ... and Lewis, D. A. (2003). Gene expression deficits in a subclass of GABA neurons in the prefrontal cortex of subjects with schizophrenia. *Journal of Neuroscience*, *23*(15), 6315-6326.

Hashimoto, T., Bazmi, H. H., Mirnics, K., Wu, Q., Sampson, A. R., and Lewis, D. A. (2008). Conserved regional patterns of GABA-related transcript expression in the neocortex of subjects with schizophrenia. *American Journal of Psychiatry*, *165*(4), 479-489.

Haug, M. F., Gesemann, M., Mueller, T., and Neuhauss, S. C. (2013). Phylogeny and expression divergence of metabotropic glutamate receptor genes in the brain of zebrafish (Danio rerio). *Journal of Comparative Neurology, 521*(7), 1533-1560.

Hawkey, A. B., Hoeng, J., Peitsch, M. C., Levin, E. D., and Koshibu, K. (2021). Subchronic effects of plant alkaloids on anxiety-like behavior in zebrafish. *Pharmacology Biochemistry and Behavior, 207*, 173223.

Hecker, A., Schulze, W., Oster, J., Richter, D. O., and Schuster, S. (2020). Removing a single neuron in a vertebrate brain forever abolishes an essential behavior. *Proceedings of the National Academy of Sciences, 117*(6), 3254-3260.

Hennah, W. and Porteous, D. (2009). The DISC1 pathway modulates expression of neurodevelopmental, synaptogenic and sensory perception genes. *PloS one, 4*(3), e4906.

Hensler, J. G. (2012). Serotonin. In *Basic neurochemistry* (pp. 300-322). Academic Press.

Higashijima, S. I., Mandel, G., and Fetcho, J. R. (2004). Distribution of prospective glutamatergic, glycinergic, and GABAergic neurons in embryonic and larval zebrafish. *Journal of Comparative Neurology, 480*(1), 1-18.

Hill, A. J., Teraoka, H., Heideman, W., and Peterson, R. E. (2005). Zebrafish as a model vertebrate for investigating chemical toxicity. *Toxicological sciences, 86*(1), 6-19.

Hinz, F. I., Aizenberg, M., Tushev, G., and Schuman, E. M. (2013). Protein synthesis-dependent associative long-term memory in larval zebrafish. *Journal of Neuroscience, 33*(39), 15382-15387.

Hodgson, S. R., Hofford, R. S., Buckman, S. G., Wellman, P. J., and Eitan, S. (2010). Morphine-induced stereotyped thigmotaxis could appear as enhanced fear and anxiety in some behavioural tests. *Journal of Psychopharmacology, 24*(6), 875-880.

Holmseth, S., Scott, H. A., Real, K., Lehre, K. P., Leergaard, T. B., Bjaalie, J. G., and Danbolt, N. C. (2009). The concentrations and distributions of three C-terminal variants of the GLT1 (EAAT2; slc1a2) glutamate transporter protein in rat brain tissue suggest differential regulation. *Neuroscience, 162*(4), 1055-1071.

Holmseth, S., Dehnes, Y., Huang, Y. H., Follin-Arbelet, V. V., Grutle, N. J., Mylonakou, M. N., ... and Danbolt, N. C. (2012). The density of EAAC1 (EAAT3) glutamate transporters expressed by neurons in the mammalian CNS. *Journal of Neuroscience, 32*(17), 6000-6013.

Hong, S., Lee, P., Baraban, S. C., and Lee, L. P. (2016). A novel long-term, multi-channel and non-invasive electrophysiology platform for zebrafish. *Scientific reports, 6*(1), 1-10.

Horacek, J., Bubenikova-Valesova, V., Kopecek, M., Palenicek, T., Dockery, C., Mohr, P., and Höschl, C. (2006). Mechanism of action of atypical antipsychotic drugs and the neurobiology of schizophrenia. *CNS drugs, 20*(5), 389-409.

Horstick, E. J., Bayleyen, Y., Sinclair, J. L., and Burgess, H. A. (2017). Search strategy is regulated by somatostatin signaling and deep brain photoreceptors in zebrafish. *BMC biology, 15*(1), 1-16.

Horstick, E. J., Bayleyen, Y., and Burgess, H. A. (2020). Molecular and cellular determinants of motor asymmetry in zebrafish. *Nature communications, 11*(1), 1-15.

Howe, D. G., Bradford, Y. M., Conlin, T., Eagle, A. E., Fashena, D., Frazer, K., ... and Westerfield, M. (2012). ZFIN, the Zebrafish Model Organism Database: increased support for mutants and transgenics. *Nucleic acids research, 41*(D1), D854-D860.

Howe, K., Clark, M. D., Torroja, C. F., Torrance, J., Berthelot, C., Muffato, M., ... and Teucke, M. (2013). The zebrafish reference genome sequence and its relationship to the human genome. *Nature, 496*(7446), 498-503.

Howes, O., McCutcheon, R., and Stone, J. (2015). Glutamate and dopamine in schizophrenia: an update for the 21st century. *Journal of psychopharmacology, 29*(2), 97-115.

Huang, Y. Y., Haug, M. F., Gesemann, M., and Neuhauss, S. C. (2012). Novel expression patterns of metabotropic glutamate receptor 6 in the zebrafish nervous system. *PLoS One, 7*(4), e35256.

Hunt, M. J., Raynaud, B., and Garcia, R. (2006). Ketamine dose-dependently induces high-frequency oscillations in the nucleus accumbens in freely moving rats. *Biological psychiatry, 60*(11), 1206-1214.

Hwang, W. Y., Fu, Y., Reyon, D., Maeder, M. L., Kaini, P., Sander, J. D., ... and Yeh, J. R. J. (2013). Heritable and precise zebrafish genome editing using a CRISPR-Cas system. *PloS one, 8*(7), e68708.

Ingebretson, J. J. and Masino, M. A. (2013). Quantification of locomotor activity in larval zebrafish: considerations for the design of high-throughput behavioral studies. *Frontiers in neural circuits, 7*, 109.

Jao, L. E., Wente, S. R., and Chen, W. (2013). Efficient multiplex biallelic zebrafish genome editing using a CRISPR nuclease system. *Proceedings of the National Academy of Sciences, 110*(34), 13904-13909.

Jetti, S. K., Vendrell-Llopis, N., and Yaksi, E. (2014). Spontaneous activity governs olfactory representations in spatially organized habenular microcircuits. *Current biology, 24*(4), 434-439.

Johnstone, M., Thomson, P. A., Hall, J., McIntosh, A. M., Lawrie, S. M., and Porteous, D. J. (2011). DISC1 in schizophrenia: genetic mouse models and human genomic imaging. *Schizophrenia bulletin, 37*(1), 14-20.

Jones, S. R., Gainetdinov, R. R., Jaber, M., Giros, B., Wightman, R. M., and Caron, M. G. (1998). Profound neuronal plasticity in response to inactivation of the dopamine transporter. *Proceedings of the National Academy of Sciences, 95*(7), 4029-4034.

Joo, W., Vivian, M. D., Graham, B. J., Soucy, E. R., and Thyme, S. B. (2021). A Customizable Low-Cost System for Massively Parallel Zebrafish Behavioral Phenotyping. *Frontiers in behavioral neuroscience, 14*, 258.

Kallai, J., Makany, T., Csatho, A., Karadi, K., Horvath, D., Kovacs-Labadi, B., ... and Jacobs, J. W. (2007). Cognitive and affective aspects of thigmotaxis strategy in humans. *Behavioral neuroscience, 121*(1), 21.

Kalueff, A. V., Gebhardt, M., Stewart, A. M., Cachat, J. M., Brimmer, M., Chawla, J. S., ... and Schneider, and the Zebrafish Neuroscience Research Consortium, H. (2013). Towards a comprehensive catalog of zebrafish behavior 1.0 and beyond. *Zebrafish, 10*(1), 70-86.

Kalueff, A. V., Stewart, A. M., and Gerlai, R. (2014). Zebrafish as an emerging model for studying complex brain disorders. *Trends in pharmacological sciences, 35*(2), 63-75.

Kalueff, A. V. (2017). Illustrated zebrafish neurobehavioral glossary. In *The rights and wrongs of zebrafish: Behavioral phenotyping of zebrafish* (pp. 291-317). Springer, Cham.

Karczmar, A. G. (1993). Brief presentation of the story and present status of studies of the vertebrate cholinergic system. *Neuropsycho-pharmacology*, *9*(3), 181-199.

Kaslin, J. A. N. and Panula, P. (2001). Comparative anatomy of the histaminergic and other aminergic systems in zebrafish (Danio rerio). *Journal of Comparative Neurology*, *440*(4), 342-377.

Kaupmann, K., Malitschek, B., Schuler, V., Heid, J., Froestl, W., Beck, P., ... and Bettler, B. (1998). GABA B-receptor subtypes assemble into functional heteromeric complexes. *Nature*, *396*(6712), 683-687.

Kettleborough, R. N., Busch-Nentwich, E. M., Harvey, S. A., Dooley, C. M., de Bruijn, E., van Eeden, F., ... and Stemple, D. L. (2013). A systematic genome-wide analysis of zebrafish protein-coding gene function. *Nature*, *496*(7446), 494-497.

Kim, D. H., Kim, J., Marques, J. C., Grama, A., Hildebrand, D. G., Gu, W., ... and Robson, D. N. (2017). Pan-neuronal calcium imaging with cellular resolution in freely swimming zebrafish. *Nature methods*, *14*(11), 1107-1114.

Kohashi, T. and Oda, Y. (2008). Initiation of Mauthner-or non-Mauthner-mediated fast escape evoked by different modes of sensory input. *Journal of Neuroscience*, *28*(42), 10641-10653.

Kok, F. O., Shin, M., Ni, C. W., Gupta, A., Grosse, A. S., van Impel, A., ... and Lawson, N. D. (2015). Reverse genetic screening reveals poor correlation between morpholino-induced and mutant phenotypes in zebrafish. *Developmental cell*, *32*(1), 97-108.

Kopman, V., Laut, J., Polverino, G., and Porfiri, M. (2013). Closed-loop control of zebrafish response using a bioinspired robotic-fish in a preference test. *Journal of the Royal Society Interface*, *10*(78), 20120540.

Korpi, E. R., Gründer, G., and Lüddens, H. (2002). Drug interactions at GABAA receptors. *Progress in neurobiology*, *67*(2), 113-159.

Kristensen, A. S., Andersen, J., Jørgensen, T. N., Sørensen, L., Eriksen, J., Loland, C. J., ... and Gether, U. (2011). SLC6 neurotransmitter transporters: structure, function, and regulation. *Pharmacological reviews*, *63*(3), 585-640.

Kyzar, E., Stewart, A. M., Landsman, S., Collins, C., Gebhardt, M., Robinson, K., and Kalueff, A. V. (2013). Behavioral effects of bidirectional modulators of brain monoamines reserpine and d-amphetamine in zebrafish. *Brain research*, *1527*, 108-116.

Laan, A., Iglesias-Julios, M., and de Polavieja, G. G. (2018). Zebrafish aggression on the sub-second time scale: evidence for mutual motor coordination and multi-functional attack manoeuvres. *Royal Society open science*, *5*(8), 180679.

Landin, J., Hovey, D., Xu, B., Lagman, D., Zettergren, A., Larhammar, D., ... and Westberg, L. (2020). Oxytocin receptors regulate social preference in zebrafish. *Scientific reports*, *10*(1), 1-12.

Langova, V., Vales, K., Horka, P., and Horacek, J. (2020). The role of zebrafish and laboratory rodents in schizophrenia research. *Frontiers in Psychiatry*, *11*.

Lawson, N. D. (2016). Reverse genetics in zebrafish: mutants, morphants, and moving forward. *Trends in cell biology*, *26*(2), 77-79.

Leary, S. C., Cobine, P. A., Kaufman, B. A., Guercin, G. H., Mattman, A., Palaty, J., ... and Shoubridge, E. A. (2007). The human cytochrome c oxidase assembly factors SCO1 and SCO2 have regulatory roles in the maintenance of cellular copper homeostasis. *Cell metabolism*, *5*(1), 9-20.

Lehre, K. P., Levy, L. M., Ottersen, O. P., Storm-Mathisen, J., and Danbolt, N. C. (1995). Differential expression of two glial glutamate transporters in the rat brain: quantitative and immunocytochemical observations. *Journal of Neuroscience*, *15*(3), 1835-1853.

Léveillé, F., Gaamouch, F. E., Gouix, E., Lecocq, M., Lobner, D., Nicole, O., and Buisson, A. (2008). Neuronal viability is controlled by a functional relation between synaptic and extrasynaptic NMDA receptors. *The FASEB Journal*, *22*(12), 4258-4271.

Levin, E. D., Bencan, Z., and Cerutti, D. T. (2007). Anxiolytic effects of nicotine in zebrafish. *Physiology and behavior*, *90*(1), 54-58.

Light, S. E. and Jontes, J. D. (2019). Multiplane calcium imaging reveals disrupted development of network topology in zebrafish pcdh19 mutants. *eneuro*, *6*(3).

Lillesaar, C. (2011). The serotonergic system in fish. *Journal of chemical neuroanatomy*, *41*(4), 294-308.

Liu, Y., Dale, S., Ball, R., VanLeuven, A. J., Sornborger, A., Lauderdale, J. D., and Kner, P. (2019). Imaging neural events in zebrafish larvae with linear structured illumination light sheet fluorescence microscopy. *Neurophotonics*, *6*(1), 015009.

Lo, W. S., Lau, C. F., Xuan, Z., Chan, C. F., Feng, G. Y., He, L., ... and Xue, H. (2004). Association of SNPs and haplotypes in GABA A receptor β 2 gene with schizophrenia. *Molecular psychiatry*, *9*(6), 603-608.

Lo, W. S., Harano, M., Gawlik, M., Yu, Z., Chen, J., Pun, F. W., ... and Xue, H. (2007). GABRB2 association with schizophrenia: commonalities and differences between ethnic groups and clinical subtypes. *Biological psychiatry*, *61*(5), 653-660.

Loviglio, M. N., Arbogast, T., Jønch, A. E., Collins, S. C., Popadin, K., Bonnet, C. S., ... and Golzio, C. (2017). The immune signaling adaptor LAT contributes to the neuroanatomical phenotype of 16p11. 2 BP2-BP3 CNVs. *The American Journal of Human Genetics*, *101*(4), 564-577.

Lucki, I. (1998). The spectrum of behaviors influenced by serotonin. *Biological psychiatry*, *44*(3), 151-162.

Malhotra, A. K., Goldman, D., Mazzanti, C., Clifton, A., Breier, A., and Pickar, D. (1998). A functional serotonin transporter (5-HTT) polymorphism is associated with psychosis in neuroleptic-free schizophrenics. *Molecular Psychiatry*, *3*(4), 328-332.

Mali, P., Yang, L., Esvelt, K. M., Aach, J., Guell, M., DiCarlo, J. E., ... and Church, G. M. (2013). RNA-guided human genome engineering via Cas9. *Science*, *339*(6121), 823-826.

Marona-Lewicka, D., Nichols, C. D., and Nichols, D. E. (2011). An animal model of schizophrenia based on chronic LSD administration: old idea, new results. *Neuropharmacology*, *61*(3), 503-512.

Marshall, F. H., Jones, K. A., Kaupmann, K., and Bettler, B. (1999). GABAB receptors–the first 7TM heterodimers. *Trends in pharmacological sciences*, *20*(10), 396-399.

Marsman, A., Van Den Heuvel, M. P., Klomp, D. W., Kahn, R. S., Luijten, P. R., and Hulshoff Pol, H. E. (2013). Glutamate in schizophrenia: a focused review and meta-analysis of 1H-MRS studies. *Schizophrenia bulletin*, *39*(1), 120-129.

Martin, S. C., Heinrich, G., and Sandell, J. H. (1998). Sequence and expression of glutamic acid decarboxylase isoforms in the developing zebrafish. *Journal of Comparative Neurology, 396*(2), 253-266.

Masellis, M., Basile, V., Meltzer, H. Y., Lieberman, J. A., Sevy, S., Macciardi, F. M., ... and Kennedy, J. L. (1998). Serotonin subtype 2 receptor genes and clinical response to clozapine in schizophrenia patients. *Neuropsychopharmacology, 19*(2), 123-132.

Matsuda, K., Yoshida, M., Kawakami, K., Hibi, M., and Shimizu, T. (2017). Granule cells control recovery from classical conditioned fear responses in the zebrafish cerebellum. *Scientific reports, 7*(1), 1-13.

Maximino, C. and Herculano, A. M. (2010). A review of monoaminergic neuropsychopharmacology in zebrafish. *Zebrafish, 7*(4), 359-378.

Medan, V. and Preuss, T. (2011). Dopaminergic-induced changes in Mauthner cell excitability disrupt prepulse inhibition in the startle circuit of goldfish. *Journal of neurophysiology, 106*(6), 3195-3204.

Meltzer, H. Y., Li, Z., Kaneda, Y., and Ichikawa, J. (2003). Serotonin receptors: their key role in drugs to treat schizophrenia. *Progress in Neuro-Psychopharmacology and Biological Psychiatry, 27*(7), 1159-1172.

Mexal, S., Berger, R., Logel, J., Ross, R. G., Freedman, R., and Leonard, S. (2010). Differential regulation of α7 nicotinic receptor gene (CHRNA7) expression in schizophrenic smokers. *Journal of molecular neuroscience, 40*(1), 185-195.

Meyer, A. and Schartl, M. (1999). Gene and genome duplications in vertebrates: the one-to-four (-to-eight in fish) rule and the evolution of novel gene functions. *Current opinion in cell biology, 11*(6), 699-704.

Meyer, M., Dhamne, S. C., LaCoursiere, C. M., Tambunan, D., Poduri, A., and Rotenberg, A. (2016). Microarray noninvasive neuronal seizure recordings from intact larval zebrafish. *PLoS One, 11*(6), e0156498.

Meyer, U. and Feldon, J. (2010). Epidemiology-driven neurodevelop-mental animal models of schizophrenia. *Progress in neurobiology, 90*(3), 285-326.

Michelotti, P., Quadros, V. A., Pereira, M. E., and Rosemberg, D. B. (2018). Ketamine modulates aggressive behavior in adult zebrafish. *Neuroscience letters, 684*, 164-168.

Millar, J. K., Wilson-Annan, J. C., Anderson, S., Christie, S., Taylor, M. S., Semple, C. A., ... and Porteous, D. J. (2000). Disruption of two novel genes by a translocation co-segregating with schizophrenia. *Human molecular genetics, 9*(9), 1415-1423.

Moens, C. B., Donn, T. M., Wolf-Saxon, E. R., and Ma, T. P. (2008). Reverse genetics in zebrafish by TILLING. *Briefings in Functional Genomics and Proteomics, 7*(6), 454-459.

Mohammad-Zadeh, L. F., Moses, L., and Gwaltney-Brant, S. M. (2008). Serotonin: a review. *Journal of veterinary pharmacology and therapeutics, 31*(3), 187-199.

Mueller, T., Vernier, P., and Wullimann, M. F. (2004). The adult central nervous cholinergic system of a neurogenetic model animal, the zebrafish Danio rerio. *Brain research, 1011*(2), 156-169.

Muto, A. and Kawakami, K. (2016). Calcium imaging of neuronal activity in free-swimming larval zebrafish. In *Zebrafish* (pp. 333-341). Humana Press, New York, NY.

Mutsatsa, S. H., Joyce, E. M., Hutton, S. B., and Barnes, T. R. (2006). Relationship between insight, cognitive function, social function and symptomatology in schizophrenia. *European Archives of Psychiatry and Clinical Neuroscience*, *256*(6), 356-363.

Nabhan, J. F., Pan, H., and Lu, Q. (2010). Arrestin domain-containing protein 3 recruits the NEDD4 E3 ligase to mediate ubiquitination of the β2-adrenergic receptor. *EMBO reports*, *11*(8), 605-611.

Nachtigall, P. G., Dias, M. C., and Pinhal, D. (2014). Evolution and genomic organization of muscle microRNAs in fish genomes. *BMC evolutionary biology*, *14*(1), 1-12.

Nakashima, A., Mori, K., Suzuki, T., Kurita, H., Otani, M., Nagatsu, T., and Ota, A. (1999). Dopamine inhibition of human tyrosine hydroxylase type 1 is controlled by the specific portion in the N-terminus of the enzyme. *Journal of neurochemistry*, *72*(5), 2145-2153.

Nass, R. and Hamza, I. (2007). The nematode C. elegans as an animal model to explore toxicology in vivo: solid and axenic growth culture conditions and compound exposure parameters. *Current Protocols in Toxicology*, *31*(1), 1-9.

Nelson, N. (1998). The family of Na+/Cl− neurotransmitter transporters. *Journal of neurochemistry*, *71*(5), 1785-1803.

Nestler, E. J. (2013). The origins of molecular psychiatry. *Journal of molecular psychiatry*, *1*(1), 1-2.

Ng, J., Zhen, J., Meyer, E., Erreger, K., Li, Y., Kakar, N., ... and Kurian, M. A. (2014). Dopamine transporter deficiency syndrome: phenotypic spectrum from infancy to adulthood. *Brain*, *137*(4), 1107-1119.

Ng, M. Y., Levinson, D. F., Faraone, S. V., Suárez, B. K., DeLisi, L. E., Arinami, T., ... and Lewis, C. M. (2009). Meta-analysis of 32 genome-wide linkage studies of schizophrenia. *Molecular psychiatry*, *14*(8), 774-785.

Niciu, M. J., Kelmendi, B., and Sanacora, G. (2012). Overview of glutamatergic neurotransmission in the nervous system. *Pharmacology Biochemistry and Behavior*, *100*(4), 656-664.

Nikiforuk, A., Hołuj, M., Kos, T., and Popik, P. (2016). The effects of a 5-HT5A receptor antagonist in a ketamine-based rat model of cognitive dysfunction and the negative symptoms of schizophrenia. *Neuropharmacology*, *105*, 351-360.

Norton, W. H., Folchert, A., and Bally-Cuif, L. (2008). Comparative analysis of serotonin receptor (HTR1A/HTR1B families) and transporter (slc6a4a/b) gene expression in the zebrafish brain. *Journal of Comparative Neurology*, *511*(4), 521-542.

Ogi, A., Licitra, R., Naef, V., Marchese, M., Fronte, B., Gazzano, A., and Santorelli, F. M. (2021). Social preference tests in zebrafish: a systematic review. *Frontiers in Veterinary Science*, *7*, 1239.

Olabi, B., Ellison-Wright, I., McIntosh, A. M., Wood, S. J., Bullmore, E., and Lawrie, S. M. (2011). Are there progressive brain changes in schizophrenia? A meta-analysis of structural magnetic resonance imaging studies. *Biological psychiatry*, *70*(1), 88-96.

Oliveira, R. F., Simões, J. M., Teles, M. C., Oliveira, C. R., Becker, J. D., and Lopes, J. S. (2016). Assessment of fight outcome is needed to activate socially driven transcriptional changes in the zebrafish brain. *Proceedings of the National Academy of Sciences*, *113*(5), E654-E661.

Orger, M. B. and de Polavieja, G. G. (2017). Zebrafish behavior: opportunities and challenges. *Annual review of neuroscience*, *40*, 125-147.

Pan, L., Shah, A. N., Phelps, I. G., Doherty, D., Johnson, E. A., and Moens, C. B. (2015). Rapid identification and recovery of ENU-induced mutations with next-generation sequencing and Paired-End Low-Error analysis. *BMC genomics*, *16*(1), 1-13.

Paoletti, P., Bellone, C., and Zhou, Q. (2013). NMDA receptor subunit diversity: impact on receptor properties, synaptic plasticity and disease. *Nature Reviews Neuroscience*, *14*(6), 383-400.

Paredes-Zúñiga, S., Ormeño, F., and Allende, M. L. (2021). Triadimefon triggers circling behavior and conditioned place preference/aversion in zebrafish in a dose dependent manner. *Neurotoxicology and Teratology*, *86*, 106979.

Patwari, P., Emilsson, V., Schadt, E. E., Chutkow, W. A., Lee, S., Marsili, A., ... and Lee, R. T. (2011). The arrestin domain-containing 3 protein regulates body mass and energy expenditure. *Cell metabolism*, *14*(5), 671-683.

Pegg, A. E. (2006). Regulation of ornithine decarboxylase. *Journal of Biological Chemistry*, *281*(21), 14529-14532.

Pegg, A. E. (2016). Functions of polyamines in mammals. *Journal of Biological Chemistry*, *291*(29), 14904-14912.

Periyasamy, S., John, S., Padmavati, R., Rajendren, P., Thirunavukkarasu, P., Gratten, J., ... and Mowry, B. J. (2019). Association of schizophrenia risk with disordered niacin metabolism in an Indian genome-wide association study. *JAMA psychiatry*, *76*(10), 1026-1034.

Perry, E., Walker, M., Grace, J., and Perry, R. (1999). Acetylcholine in mind: a neurotransmitter correlate of consciousness?. *Trends in neurosciences*, *22*(6), 273-280.

Petroff, O. A. (2002). Book review: GABA and glutamate in the human brain. *The Neuroscientist*, *8*(6), 562-573.

Pham, M., Raymond, J., Hester, J., Kyzar, E., Gaikwad, S., Bruce, I., ... and Kalueff, A. V. (2012). Assessing social behavior phenotypes in adult zebrafish: Shoaling, social preference, and mirror biting tests. In *Zebrafish protocols for neurobehavioral research* (pp. 231-246). Humana Press, Totowa, NJ.

Phillips, J. B. and Westerfield, M. (2014). Zebrafish models in translational research: tipping the scales toward advancements in human health. *Disease models and mechanisms*, *7*(7), 739-743.

Pineda, R., Beattie, C. E., and Hall, C. W. (2011). Recording the adult zebrafish cerebral field potential during pentylenetetrazole seizures. *Journal of neuroscience methods*, *200*(1), 20-28.

Pithadia, A. B. and Jain, S. M. (2009). 5-Hydroxytryptamine receptor subtypes and their modulators with therapeutic potentials. *Journal of clinical medicine research*, *1*(2), 72.

Png, W. Y., TANG, P., Ogawa, S., Parhar, I., and Mok, S. Y. (2021). Startle Habituation: A Tool for Assessing Information Processing Deficits in Zebrafish Model of Schizophrenia. *Sains Malaysiana*, *50*(1), 201-206.

Polverino, G. and Porfiri, M. (2013). Zebrafish (Danio rerio) behavioural response to bioinspired robotic fish and mosquitofish (Gambusia affinis). *Bioinspiration and biomimetics*, *8*(4), 044001.

Porcelli, S., Van Der Wee, N., van der Werff, S., Aghajani, M., Glennon, J. C., van Heukelum, S., ... and Serretti, A. (2019). Social brain, social dysfunction and social withdrawal. *Neuroscience and Biobehavioral Reviews*, *97*, 10-33.

Porteous, D. J., Millar, J. K., Brandon, N. J., and Sawa, A. (2011). DISC1 at 10: connecting psychiatric genetics and neuroscience. *Trends in molecular medicine*, *17*(12), 699-706.

Portugues, R., Feierstein, C. E., Engert, F., and Orger, M. B. (2014). Whole-brain activity maps reveal stereotyped, distributed networks for visuomotor behavior. *Neuron*, *81*(6), 1328-1343.

Postlethwait, J., Amores, A., Force, A., and Yan, Y. L. (1998). The zebrafish genome. *Methods in cell biology*, *60*, 149-163.

Poulin, J. F., Caronia, G., Hofer, C., Cui, Q., Helm, B., Ramakrishnan, C., ... and Awatramani, R. (2018). Mapping projections of molecularly defined dopamine neuron subtypes using intersectional genetic approaches. *Nature neuroscience*, *21*(9), 1260-1271.

Pow, D. V. and Barnett, N. L. (2000). Developmental expression of excitatory amino acid transporter 5: a photoreceptor and bipolar cell glutamate transporter in rat retina. *Neuroscience letters*, *280*(1), 21-24.

Powell, C. M. and Miyakawa, T. (2006). Schizophrenia-relevant behavioral testing in rodent models: a uniquely human disorder?. *Biological psychiatry*, *59*(12), 1198-1207.

Prut, L. and Belzung, C. (2003). The open field as a paradigm to measure the effects of drugs on anxiety-like behaviors: a review. *European journal of pharmacology*, *463*(1-3), 3-33.

Pun, F. W., Zhao, C., Lo, W. S., Ng, S. K., Tsang, S. Y., Nimgaonkar, V., ... and Xue, H. (2011). Imprinting in the schizophrenia candidate gene GABRB2 encoding GABA A receptor β 2 subunit. *Molecular psychiatry*, *16*(5), 557-568.

Raedler, T. J. and Tandon, R. (2006). Cholinergic mechanisms in schizophrenia: Current concepts. *Current Psychosis and Therapeutics Reports*, *4*(1), 20-26.

Randlett, O., Wee, C. L., Naumann, E. A., Nnaemeka, O., Schoppik, D., Fitzgerald, J. E., ... and Schier, A. F. (2015). Whole-brain activity mapping onto a zebrafish brain atlas. *Nature methods*, *12*(11), 1039-1046.

Rico, E. P., de Oliveira, D. L., Rosemberg, D. B., Mussulini, B. H., Bonan, C. D., Dias, R. D., ... and Bogo, M. R. (2010). Expression and functional analysis of Na+-dependent glutamate transporters from zebrafish brain. *Brain research bulletin*, *81*(4-5), 517-523.

Rink, E. and Wullimann, M. F. (2001). The teleostean (zebrafish) dopaminergic system ascending to the subpallium (striatum) is located in the basal diencephalon (posterior tuberculum). *Brain research*, *889*(1-2), 316-330.

Rink, E. and Wullimann, M. F. (2002). Connections of the ventral telencephalon and tyrosine hydroxylase distribution in the zebrafish brain (Danio rerio) lead to identification of an ascending dopaminergic system in a teleost. *Brain research bulletin*, *57*(3-4), 385-387.

Ritsner, M. S. and Blumenkrantz, H. (2007). Predicting domain-specific insight of schizophrenia patients from symptomatology, multiple neurocognitive functions, and personality related traits. *Psychiatry research, 149*(1-3), 59-69.

Roberts, A. C., Bill, B. R., and Glanzman, D. L. (2013). Learning and memory in zebrafish larvae. *Frontiers in neural circuits, 7*, 126.

Romero-Ferrero, F., Bergomi, M. G., Hinz, R. C., Heras, F. J., and de Polavieja, G. G. (2019). Idtracker. ai: tracking all individuals in small or large collectives of unmarked animals. *Nature methods, 16*(2), 179-182.

Ruberto, T., Mwaffo, V., Singh, S., Neri, D., and Porfiri, M. (2016). Zebrafish response to a robotic replica in three dimensions. *Royal Society open science, 3*(10), 160505.

Rudolph, U. and Möhler, H. (2004). Analysis of GABAA receptor function and dissection of the pharmacology of benzodiazepines and general anesthetics through mouse genetics. *Annu. Rev. Pharmacol. Toxicol., 44*, 475-498.

Rudolph, U. and Möhler, H. (2006). GABA-based therapeutic approaches: GABAA receptor subtype functions. *Current opinion in pharmacology, 6*(1), 18-23.

Rujescu, D. (2017). Search for risk genes in schizophrenia. *Der Nervenarzt, 88*(7), 751-754.

Sackerman, J., Donegan, J. J., Cunningham, C. S., Nguyen, N. N., Lawless, K., Long, A., ... and Gould, G. G. (2010). Zebrafish behavior in novel environments: effects of acute exposure to anxiolytic compounds and choice of Danio rerio line. *International journal of comparative psychology/ISCP; sponsored by the International Society for Comparative Psychology and the University of Calabria, 23*(1), 43.

Sakai, C., Ijaz, S., and Hoffman, E. J. (2018). Zebrafish models of neurodevelopmental disorders: past, present, and future. *Frontiers in molecular neuroscience, 11*, 294.

Sallinen, V., Sundvik, M., Reenilä, I., Peitsaro, N., Khrustalyov, D., Anichtchik, O., ... and Panula, P. (2009). Hyperserotonergic phenotype after monoamine oxidase inhibition in larval zebrafish. *Journal of neurochemistry, 109*(2), 403-415.

Sander, J. D., Cade, L., Khayter, C., Reyon, D., Peterson, R. T., Joung, J. K., and Yeh, J. R. J. (2011). Targeted gene disruption in somatic zebrafish cells using engineered TALENs. *Nature biotechnology, 29*(8), 697-698.

Sanger Institute Zebrafish Mutation Project (2013) ZFIN Feature: sa17298. https://zfin.org/ZDB-ALT-131217-13434.

Santiago, M. P. and Potter, L. T. (2001). Biotinylated m4-toxin demonstrates more M4 muscarinic receptor protein on direct than indirect striatal projection neurons. *Brain research, 894*(1), 12-20.

Schäfer, M. K. H., Varoqui, H., Defamie, N., Weihe, E., and Erickson, J. D. (2002). Molecular cloning and functional identification of mouse vesicular glutamate transporter 3 and its expression in subsets of novel excitatory neurons. *Journal of Biological Chemistry, 277*(52), 50734-50748.

Schauerte, H. E., Van Eeden, F. J., Fricke, C., Odenthal, J., Strahle, U., and Haffter, P. (1998). Sonic hedgehog is not required for the induction of medial floor plate cells in the zebrafish. *Development, 125*(15), 2983-2993.

Schleimer, S. B., Hinton, T., Dixon, G., and Johnston, G. A. (2004). GABA transporters GAT-1 and GAT-3 in the human dorsolateral prefrontal cortex in schizophrenia. *Neuropsychobiology, 50*(3), 226-230.

Schneider, H., Fritzky, L., Williams, J., Heumann, C., Yochum, M., Pattar, K., ... and Hawley, E. (2012). Cloning and expression of a zebrafish 5-HT2C receptor gene. *Gene, 502*(2), 108-117.

Schnörr, S. J., Steenbergen, P. J., Richardson, M. K., and Champagne, D. L. (2012). Measuring thigmotaxis in larval zebrafish. *Behavioural brain research, 228*(2), 367-374.

Schousboe, A., Madsen, K. K., Barker-Haliski, M. L., and White, H. S. (2014). The GABA synapse as a target for antiepileptic drugs: a historical overview focused on GABA transporters. *Neurochemical research, 39*(10), 1980-1987.

Schroeder, P., Jones, S., Young, I. S., and Sneddon, L. U. (2014). What do zebrafish want? Impact of social grouping, dominance and gender on preference for enrichment. *Laboratory animals, 48*(4), 328-337.

Scimemi, A. (2014). Structure, function, and plasticity of GABA transporters. *Frontiers in cellular neuroscience, 8*, 161.

Seeman, P. (2014). Clozapine, a fast-off-D2 antipsychotic. *ACS chemical neuroscience, 5*(1), 24-29.

Severinsen, K., Sinning, S., Müller, H. K., and Wiborg, O. (2008). Characterisation of the zebrafish serotonin transporter functionally links TM10 to the ligand binding site. *Journal of neurochemistry, 105*(5), 1794-1805.

Shimada, S., Kitayama, S., Walther, D., and Uhl, G. (1992). Dopamine transporter mRNA: dense expression in ventral midbrain neurons. *Molecular brain research, 13*(4), 359-362.

Singh, S. P., Singh, V., Kar, N., and Chan, K. (2010). Efficacy of antidepressants in treating the negative symptoms of chronic schizophrenia: meta-analysis. *The British Journal of Psychiatry, 197*(3), 174-179.

Sison, M. and Gerlai, R. (2010). Associative learning in zebrafish (Danio rerio) in the plus maze. *Behavioural brain research, 207*(1), 99-104.

Soghomonian, J. J. and Martin, D. L. (1998). Two isoforms of glutamate decarboxylase: why?. *Trends in pharmacological sciences, 19*(12), 500-505.

Souza, B. R., Romano-Silva, M. A., and Tropepe, V. (2011). Dopamine D2 receptor activity modulates Akt signaling and alters GABAergic neuron development and motor behavior in zebrafish larvae. *Journal of Neuroscience, 31*(14), 5512-5525.

Spinello, C., Yang, Y., Macrì, S., and Porfiri, M. (2019). Zebrafish adjust their behavior in response to an interactive robotic predator. *Frontiers in Robotics and AI, 6*, 38.

Sprague, J., Clements, D., Conlin, T., Edwards, P., Frazer, K., Schaper, K., ... and Westerfield, M. (2003). The Zebrafish Information Network (ZFIN): the zebrafish model organism database. *Nucleic acids research, 31*(1), 241-243.

Stewart, A. M., Braubach, O., Spitsbergen, J., Gerlai, R., and Kalueff, A. V. (2014). Zebrafish models for translational neuroscience research: from tank to bedside. *Trends in neurosciences, 37*(5), 264-278.

Sturgeon, R. D., Fessler, R. G., and Meltzer, H. Y. (1979). Behavioral rating scales for assessing phencyclidine-induced locomotor activity, stereotypes behavior and ataxia in rats. *European journal of pharmacology, 59*(3-4), 169-179.

Sullivan, P. F., Kendler, K. S., and Neale, M. C. (2003). Schizophrenia as a complex trait: evidence from a meta-analysis of twin studies. *Archives of general psychiatry*, *60*(12), 1187-1192.

Sun, M., Li, W., Jiao, Z., and Zhao, X. (2019, July). A multi-target tracking platform for zebrafish based on deep neural network. In *2019 IEEE 9th Annual International Conference on CYBER Technology in Automation, Control, and Intelligent Systems (CYBER)* (pp. 637-642). IEEE.

Sundvik, M. and Panula, P. (2012). Organization of the histaminergic system in adult zebrafish (Danio rerio) brain: neuron number, location, and cotransmitters. *Journal of Comparative Neurology*, *520*(17), 3827-3845.

Swain, H. A., Sigstad, C., and Scalzo, F. M. (2004). Effects of dizocilpine (MK-801) on circling behavior, swimming activity, and place preference in zebrafish (Danio rerio). *Neurotoxicology and teratology*, *26*(6), 725-729.

Takahashi, M., Narushima, M., and Oda, Y. (2002). In vivo imaging of functional inhibitory networks on the mauthner cell of larval zebrafish. *Journal of Neuroscience*, *22*(10), 3929-3938.

Takahashi, M., Inoue, M., Tanimoto, M., Kohashi, T., and Oda, Y. (2017). Short-term desensitization of fast escape behavior associated with suppression of Mauthner cell activity in larval zebrafish. *Neuroscience research*, *121*, 29-36.

Takamori, S., Malherbe, P., Broger, C., and Jahn, R. (2002). Molecular cloning and functional characterization of human vesicular glutamate transporter 3. *EMBO reports*, *3*(8), 798-803.

Tandon, R. and Fleischhacker, W. W. (2005). Comparative efficacy of antipsychotics in the treatment of schizophrenia: a critical assessment. *Schizophrenia Research*, *79*(2-3), 145-155.

Tang, W., Davidson, J. D., Zhang, G., Conen, K. E., Fang, J., Serluca, F., ... and Fishman, M. C. (2020). Genetic control of collective behavior in zebrafish. *IScience*, *23*(3), 100942.

Tarabeux, J., Champagne, N., Brustein, E., Hamdan, F. F., Gauthier, J., Lapointe, M., ... and Rouleau, G. A. (2010). De novo truncating mutation in Kinesin 17 associated with schizophrenia. *Biological psychiatry*, *68*(7), 649-656.

Teame, T., Zhang, Z., Ran, C., Zhang, H., Yang, Y., Ding, Q., ... and Zhou, Z. (2019). The use of zebrafish (Danio rerio) as biomedical models. *Animal Frontiers*, *9*(3), 68-77.

Teles, M. C., Dahlbom, S. J., Winberg, S., and Oliveira, R. F. (2013). Social modulation of brain monoamine levels in zebrafish. *Behavioural brain research*, *253*, 17-24.

Teles, M. C., Gozdowska, M., Kalamarz-Kubiak, H., Kulczykowska, E., and Oliveira, R. F. (2016). Agonistic interactions elicit rapid changes in brain nonapeptide levels in zebrafish. *Hormones and behavior*, *84*, 57-63.

Thompson, M., Weickert, C. S., Wyatt, E., and Webster, M. J. (2009). Decreased glutamic acid decarboxylase67 mRNA expression in multiple brain areas of patients with schizophrenia and mood disorders. *Journal of psychiatric research*, *43*(11), 970-977.

Thyme, S. B., Pieper, L. M., Li, E. H., Pandey, S., Wang, Y., Morris, N. S., ... and Schier, A. F. (2019). Phenotypic landscape of schizophrenia-associated genes defines candidates and their shared functions. *Cell*, *177*(2), 478-491.

Tian, N., Petersen, C., Kash, S., Baekkeskov, S., Copenhagen, D., and Nicoll, R. (1999). The role of the synthetic enzyme GAD65 in the control of neuronal γ-aminobutyric acid release. *Proceedings of the National Academy of Sciences*, *96*(22), 12911-12916.

Tolwinski, N. S. (2017). Introduction: Drosophila—a model system for developmental biology.

Tomasello, D. L. and Sive, H. (2020). Noninvasive Multielectrode Array for Brain and Spinal Cord Local Field Potential Recordings from Live Zebrafish Larvae. *Zebrafish*, *17*(4), 271-277.

Tomoda, T., Sumitomo, A., Jaaro-Peled, H., and Sawa, A. (2016). Utility and validity of DISC1 mouse models in biological psychiatry. *Neuroscience*, *321*, 99-107.

Tran, S., Muraleetharan, A., Fulcher, N., Chatterjee, D., and Gerlai, R. (2016). MK-801 increases locomotor activity in a context-dependent manner in zebrafish. *Behavioural brain research*, *296*, 26-29.

Tunbak, H., Vazquez-Prada, M., Ryan, T. M., Kampff, A. R., and Dreosti, E. (2020). Whole-brain mapping of socially isolated zebrafish reveals that lonely fish are not loners. *Elife*, *9*, e55863.

Van Os, J., Rutten, B. P., and Poulton, R. (2008). Gene-environment interactions in schizophrenia: review of epidemiological findings and future directions. *Schizophrenia bulletin*, *34*(6), 1066-1082.

Varshney, G. K., Lu, J., Gildea, D. E., Huang, H., Pei, W., Yang, Z., ... and Lin, S. (2013). A large-scale zebrafish gene knockout resource for the genome-wide study of gene function. *Genome research*, *23*(4), 727-735.

Vecchia, E. D., Donato, V. D., Young, A. M., Bene, F. D., & Norton, W. H. (2019). Reelin Signaling controls the preference for social novelty in Zebrafish. *Frontiers in Behavioral Neuroscience*, *13*, 214-214.

Vendrell-Llopis, N. and Yaksi, E. (2015). Evolutionary conserved brainstem circuits encode category, concentration and mixtures of taste. *Scientific reports*, *5*(1), 1-10.

Viana, J., Wildman, N., Hannon, E., Farbos, A., O'Neill, P., Moore, K., ... and Mill, J. (2020). Clozapine-induced transcriptional changes in the zebrafish brain. *NPJ schizophrenia*, *6*(1), 1-11.

Vidović, B., Đorđević, B., Milovanović, S., Škrivanj, S., Pavlović, Z., Stefanović, A., and Kotur-Stevuljević, J. (2013). Selenium, zinc, and copper plasma levels in patients with schizophrenia: relationship with metabolic risk factors. *Biological trace element research*, *156*(1), 22-28.

Vithlani, M., Terunuma, M., and Moss, S. J. (2011). The dynamic modulation of GABAA receptor trafficking and its role in regulating the plasticity of inhibitory synapses. *Physiological reviews*, *91*(3), 1009-1022.

Volk, D. W., Austin, M. C., Pierri, J. N., Sampson, A. R., and Lewis, D. A. (2001). GABA transporter-1 mRNA in the prefrontal cortex in schizophrenia: decreased expression in a subset of neurons. *American Journal of Psychiatry*, *158*(2), 256-265.

Walker, C. and Streisinger, G. (1983). Induction of mutations by γ-rays in pregonial germ cells of zebrafish embryos. *Genetics*, *103*(1), 125-136.

Walther, D. J. and Bader, M. (2003). A unique central tryptophan hydroxylase isoform. *Biochemical pharmacology*, *66*(9), 1673-1680.

Walz, N., Mühlberger, A., and Pauli, P. (2016). A human open field test reveals thigmotaxis related to agoraphobic fear. *Biological psychiatry*, *80*(5), 390-397.

Wang, G., Zhang, G., Li, Z., Fawcett, C. H., Coble, M., Sosa, M. X., ... and Fishman, M. C. (2019). Abnormal behavior of zebrafish mutant in dopamine transporter is rescued by clozapine. *Iscience*, *17*, 325-333.

Wang, L., Jiang, W., Lin, Q., Zhang, Y., and Zhao, C. (2016). DNA methylation regulates gabrb2 mRNA expression: developmental variations and disruptions in l-methionine-induced zebrafish with schizophrenia-like symptoms. *Genes, Brain and Behavior*, *15*(8), 702-710.

Way, G. P., Ruhl, N., Snekser, J. L., Kiesel, A. L., and McRobert, S. P. (2015). A comparison of methodologies to test aggression in zebrafish. *Zebrafish*, *12*(2), 144-151.

Wess, J., Blin, N., Mutschler, E., and Blüml, K. (1995). Muscarinic acetylcholine receptors: structural basis of ligand binding and G protein coupling. *Life sciences*, *56*(11-12), 915-922.

Wess, J. (1996). Molecular biology of muscarinic acetylcholine receptors. *Critical Reviews™ in Neurobiology*, *10*(1).

Wess, J., Duttaroy, A., Zhang, W., Gomeza, J., Cui, Y., Miyakawa, T., ... and Yamada, M. (2003). M 1-M 5 muscarinic receptor knockout mice as novel tools to study the physiological roles of the muscarinic cholinergic system. *Receptors and channels*, *9*(4), 279-290.

Wess, J. (2004). Muscarinic acetylcholine receptor knockout mice: novel phenotypes and clinical implications. *Annu. Rev. Pharmacol. Toxicol.*, *44*, 423-450.

White, R. M. (2015). Cross-species oncogenomics using zebrafish models of cancer. *Current opinion in genetics and development*, *30*, 73-79.

Whiting, P. J. (2006). GABA-A receptors: a viable target for novel anxiolytics?. *Current opinion in pharmacology*, *6*(1), 24-29.

Wienholds, E., Schulte-Merker, S., Walderich, B., and Plasterk, R. H. (2002). Target-selected inactivation of the zebrafish rag1 gene. *Science*, *297*(5578), 99-102.

Wienholds, E., Van Eeden, F., Kosters, M., Mudde, J., Plasterk, R. H., and Cuppen, E. (2003). Efficient target-selected mutagenesis in zebrafish. *Genome research*, *13*(12), 2700-2707.

Wierońska, J. M., Kusek, M., Tokarski, K., Wabno, J., Froestl, W., and Pilc, A. (2011). The GABAB receptor agonist CGP44532 and the positive modulator GS39783 reverse some behavioural changes related to positive syndromes of psychosis in mice. *British journal of pharmacology*, *163*(5), 1034-1047.

Wood, J. D., Bonath, F., Kumar, S., Ross, C. A., and Cunliffe, V. T. (2009). Disrupted-in-schizophrenia 1 and neuregulin 1 are required for the specification of oligodendrocytes and neurones in the zebrafish brain. *Human molecular genetics*, *18*(3), 391-404.

Woolf, N. J. (1991). Cholinergic systems in mammalian brain and spinal cord. *Progress in neurobiology*, *37*(6), 475-524.

Xie, J., Jusuf, P. R., Bui, B. V., Dudczig, S., Sztal, T. E., and Goodbourn, P. T. (2021). Altered visual function in a larval zebrafish knockout of neurodevelopmental risk gene pdzk1. *Investigative ophthalmology and visual science*, *62*(3), 29-29.

Xu, J., Kurup, P., Zhang, Y., Goebel-Goody, S. M., Wu, P. H., Hawasli, A. H., ... and Lombroso, P. J. (2009). Extrasynaptic NMDA receptors couple preferentially to excitotoxicity via calpain-mediated cleavage of STEP. *Journal of Neuroscience, 29*(29), 9330-9343.

Yang, P., Takahashi, H., Murase, M., and Itoh, M. (2021). Zebrafish behavior feature recognition using three-dimensional tracking and machine learning. *Scientific reports, 11*(1), 1-9.

Zabegalov, K. N., Kolesnikova, T. O., Khatsko, S. L., Volgin, A. D., Yakovlev, O. A., Amstislavskaya, T. G., ... and Kalueff, A. V. (2019a). Understanding zebrafish aggressive behavior. *Behavioural processes, 158*, 200-210.

Zabegalov, K. N., Khatsko, S. L., Lakstygal, A. M., Demin, K. A., Cleal, M., Fontana, B. D., ... and Kalueff, A. V. (2019b). Abnormal repetitive behaviors in zebrafish and their relevance to human brain disorders. *Behavioural brain research, 367*, 101-110.

Zdebik, A. A., Mahmood, F., Stanescu, H. C., Kleta, R., Bockenhauer, D., and Russell, C. (2013). Epilepsy in kcnj10 morphant zebrafish assessed with a novel method for long-term EEG recordings. *PloS one, 8*(11), e79765.

Zhang, S., Liu, X., Sun, M., Zhang, Q., Li, T., Li, X., ... and Feng, X. (2018). Reversal of reserpine-induced depression and cognitive disorder in zebrafish by sertraline and Traditional Chinese Medicine (TCM). *Behavioral and Brain Functions, 14*(1), 1-14.

Zhao, C., Xu, Z., Chen, J., Yu, Z., Tong, K. L., Lo, W. S., ... and Xue, H. (2006). Two isoforms of GABA A receptor β 2 subunit with different electrophysiological properties: differential expression and genotypical correlations in schizophrenia. *Molecular psychiatry, 11*(12), 1092-1105.

Zhao, C., Xu, Z., Wang, F., Chen, J., Ng, S. K., Wong, P. W., ... and Xue, H. (2009). Alternative-splicing in the exon-10 region of GABAA receptor β2 subunit gene: relationships between novel isoforms and psychotic disorders. *PloS one, 4*(9), e6977.

Zhao, C., Wang, F., Pun, F. W., Mei, L., Ren, L., Yu, Z., ... and Xue, H. (2012). Epigenetic regulation on GABRB2 isoforms expression: developmental variations and disruptions in psychotic disorders. *Schizophrenia research, 134*(2-3), 260-266.

Zhou, Y. and Danbolt, N. C. (2013). GABA and glutamate transporters in brain. *Frontiers in endocrinology, 4*, 165.

Zhuo, C., Hou, W., Hu, L., Lin, C., Chen, C., and Lin, X. (2017). Genomic editing of non-coding RNA genes with CRISPR/Cas9 ushers in a potential novel approach to study and treat schizophrenia. *Frontiers in molecular neuroscience, 10*, 28.

Chapter 2

Melatonin: Pleiotropic Action for the Regulation of Fish Reproduction in Aquaculture

Joydeep Das, Akash Acharyya and Kazi Nurul Hasan*

Department of Zoology, Sidho Kanho Birsha University, Purulia, India

Abstract

Melatonin (*N*-acetyl-5-methoxy-tryptamine), a naturally occurring chronobiotic molecule, is principally produced in the pineal organ in response to ambient light-dark (LD) cycle in a rhythmic fashion. Additionally, it is also synthesized from various extra-pineal sources in fish like brain, retina, gut, liver, ovary etc. This indoleamine is bio-synthesized from aromatic amino acid L-Tryptophan through four enzyme-catalyzed reactions, and arylalkylamine *N*-acetyltransferase (AANAT), converts serotonin to *N*-acetyl serotonin, is the rate-limiting enzyme of this pathway. Presence of this ~23 kDa key regulator enzyme in fish pineal, retina, gut, liver is studied elaborately but till date the data on endogenous synthesis of melatonin in reproductive organs are not so sufficient. In last few years, many researchers explored the regulatory pathway of melatonergic system in male and female reproduction and they also hypothesized that this hormone may be applied in aquaculture system for better production of fish to fulfill its demand as table food. Melatonin not only performs a multitude of regulatory activities in seasonal reproduction, it has also a wide range of pleiotropic actions to control the other physiological functions in aquaculture like digestive

* Corresponding Author's Email: kazihasan5@gmail.com.

In: Advances in Animal Science and Zoology. Volume 20
Editor: Owen P. Jenkins
ISBN: 979-8-88697-199-6

system, immune system, etc. This pineal hormone acts as a primary synchronizer between environment and the neuroendocrine system for the regulation of seasonal reproduction in vertebrates. However, seasonal breeders rely on their environment to regulate various biological rhythms that control physiological processes such as gonadal maturity, and melatonin, aids in the regulation of this progressions to ensure efficacious reproduction and maximum survival of the offspring. The annual gonadal cycle in majority of seasonally reproducing fish consists of: i) preparatory phase, in which germ cells undergoes gametogenic recrudescence; ii) a long pre-spawning phase, where cells undergo growth and development; iii) a brief spawning phase during which gonadal cells complete their maturation; and iv) post-spawning phase where regression of cells takes place. The reproductive phases in an annual gonadal cycle appear to be markedly under the control of an endogenous clock framework that offers vital information about the "time of year" and changing environmental signals in a species-specific way. All the 'clock-controlled genes' (CCGs) are primarily regulated by the endogenous circadian clock, which is located in the suprachiasmatic nucleus (SCN) of the hypothalamus and in other peripheral organs as well. Expression of *aanat* gene is a crucial switch in melatonin production and is regulated as a CCG. Clock genes and a complex 'transcriptional and translational feedback loop' (TTFL) govern reproductive cycle, but the involvement of melatonin and the mechanism by which it controls rhythmic molecular events, remain a mystery. Melatonin regulates gonadal development *via* hypothalamus-hypophyseal-gonadal (HPG) axis in a specific reproductive phase dependent manner. Substantial data gathered from a number of carefully controlled studies employing exogenous melatonin administration and/or altered photoperiods (long- and short photoperiods) to several fish species, especially in carp, concluded that it has stimulatory or inhibitory or no effects at all on the functions of gonads depending on the reproductive status of concerned fish. Recent studies on carp also demonstrated that ovarian melatonin concentration exhibited seasonal variations with titer reaching a peak during the spawning season and then considerably declining all through the post-spawning phase of the annual reproductive cycle. Melatonin may potentially have a function in regulating vitellogenesis in liver, and the incorporation of vitellogenin protein in oocytes (a marker of advancement of germ cells growth and maturation) as well, according to the evidences. Another important contribution from the study on carp ovary clearly suggested that melatonin accelerates the maturation of ovaprim-(synthetic GnRH and domperidone) induced oocytes by activating the structural components of maturation promoting factor (MPF) i.e., cyclin B and Cdc2; and this study also hypothesized that the hormonal combination may be used as "biological tool" in aquaculture for successful spawning. However, detection of high affinity melatonin binding sites (~37 kDa MT1 receptor protein) in carp oocytes argued in favor of receptor mediated action of melatonin in reproduction. Interestingly, due to lipophilic nature of melatonin, it also permeates the plasma membrane in living cells and acts as a potent antioxidant in a receptor independent pathway. It aids in the diminution of oxidative stress by scavenging oxygen centered free radicals directly and also by increasing the activities of several antioxidative enzymes and minimize the ovulatory oxidative stress during spawning phase in fish. A better knowledge on the reproductive physiology of seasonal breeders such as fish may be gained by serious attention on the intricate network of clock genes and their associative interaction with melatonergic system on the regulation of these events. However, despite a great deal of research in the past few decades and current studies, involvement of melatonin in fish reproduction, especially in male, and its potential application in aquaculture is still an enigma. The purpose of this chapter is to consolidate the required facts about melatonin and its involvement in the regulation of fish reproduction in order to provide relevant information for subsequent investigations.

Keywords: aquaculture, fish, melatonin, reproduction, stress, clock genes

1. Introduction

Melatonin (*N*-acetyl 5-methoxytryptamine) was first isolated, purified and characterized from the extracts of bovine pineal gland by an American dermatologist Dr. Aaron Bunsen Lerner at the Yale University School of Medicine in 1958 (Lerner et al., 1958). The background of the melatonin discovery was started at early 1900s when two scientists Carey Pratt McCord and Floyd P. Allen shared their novel observations on the effects of pineal extracts on skin pigmentation in amphibia (McCord and Allen, 1917). They observed for the first time that applications of crude acetone extracts of bovine pineal glands caused over-blanching of skin which resulted into the clear visibility of larger viscera through the dorsal body wall in the tadpole of *Rana pipens*. After 40 years of this novel observation, Lerner and his co-workers named this pineal indole substance as 'melatonin' for its aggregative effects on 'melanin pigment granules' in accordance to the initial findings of McCord and Allen's work (McCord and Allen, 1917; Lerner et al., 1958) (Figure 1). Melatonin has a wide range of pleiotropic physiological and therapeutic potentials in all living systems ranging from bacteria to unicellular eukaryotes, invertebrate to higher vertebrate animals and is distributed ubiquitously in various tissues and organs in most of the studied vertebrates (Chowdhury and Maitra, 2012). This hormone is synthesized from amino acid L-tryptophan through four enzyme catalyzed reactions and primarily, it is rhythmically produced within the pinealocytes in vertebrates (Maitra et al., 2005), and this endogenous production as well as secretion is well synchronized with the environmental light-dark cycle (Falcón et al., 2010). It is also detected in different extra-pineal sources including retina (Iuvone and Besharse, 1983), Harderian gland (Menendez-Pelaez et al., 1987), extraorbital lacrimal or Lowenthal's gland (Mhatre et al., 1988), gastrointestinal (GI) tract (Bubenik, 1980), liver (Hasan et al., 2016), gall bladder (Bubenik, 2002), gonads (Itoh et al., 1999; Hasan et al., 2014) in different animal groups ranging from lower vertebrates to mammals. Melatonin is considered to be an evolutionary conserved molecule across different living systems studied so far and its maximum production is observed during night or dark phase regardless of the diurnal or nocturnal activity of the organism (Falcón et al., 2010; Maitra et al., 2015). Considering this conserved feature of melatonin, it is regarded as a "chronobiotic" substance which is involved in giving information of the vital environmental cues such as the length of the night or time of the day or the time of the year to all the tissues and cells of the organisms (Pevet et al., 2006; Falcón et al., 2011). This unique feature of melatonin helps to regulate various rhythmic physiological functions especially seasonal reproduction in a broad range of vertebrates, including fish (Amano et al., 2000; Falcón et al., 2010). However, the regulatory role of melatonin on seasonal reproduction of fish is well studied in a tropical teleost *Catla catla* (Maitra et al., 2015) but the major concern about the importance of this hormone in aquaculture has not yet been duly addressed in any group of fish. Reproductive cycle in most of the fish is seasonal and the periodic changes in various environmental factors such as photoperiod, temperature, rainfall etc. play a very crucial role in the proper synchronization with reproduction in aquaculture (Maitra, 2011). To fulfill our nutritional demands, increase in the production of fish as a chief source of protein is getting major attention many folds for last few decades in aquaculture globally (Acharyya et al., 2021).

Reproductive events in an annual gonadal cycle in most of the fish species consist of a very short period of peak reproductively active or "spawning," which is preceded by a long and complicated process of "preparation." Temporal organization of seasonal reproduction may be the function of a self-sustained circannual rhythm (Duston and Bromage, 1986) or, influenced by one or more components of the environment (Bromage et al., 2001). Specifically, different limnological as well as physico-chemical parameters are well synchronized with the gonadal cycle in such a manner that: 1) initiation of gametogenesis during "preparation" takes place so that at the time of spawning abundant mature healthy gametes will be available (Bromage et al., 2001; Servili et al., 2020); 2) spawning takes place in the specific suitable part of the year to ensure maximum survival of the progeny (Foster and Kreitzman, 2005). This synchronization is established by the synthesis of melatonin from major circadian oscillator, the pineal organ in large variety of temperate and in various air-breathing fish species (Falcón et al., 2010; Maitra et al., 2013; Sánchez-Vázquez et al., 2019). The effects of melatonin in the control of gonadal growth and maturation in fish have been investigated either by pinealectomy (Popek et al., 1997), or by melatonin administration (Moniruzzaman et al., 2016) or both (Joy and Khan, 1991). Notably, exogenous melatonin administration induces variable influences like, pro-gonadal or, anti-gonadal or, no subtle effects on gonads, depending upon the reproductive status of fish in an annual cycle (Maitra et al., 2005; Bhattacharya et al., 2007). However, the mechanism of action is now well established from different carefully controlled experiments that melatonin exerts its actions in the regulation of gametogenesis in fish reproduction by interacting with various hormones (GnRH, GnIH, Kisspeptin, GTH, sex steroids) present in hypothalamo-hypophyseal-gonadal (HPG) axis (Falcón et al., 2010; Pal et al., 2012; Tsutsui and Ubuka, 2020, Hafeez and Ahmad, 2021). Melatonin modulates the functions of sex steroids (17β-estradiol, 17α, 20β-dihydroxy progesterone) during oocyte growth and maturation by its direct actions on the gonads through specific membrane bound G-protein coupled receptors (MT1 and MT2) on the target site in fish species (Maitra and Hasan, 2016). Additionally, the presence of endogenous melatonin within carp ovary also indicates its role in the development of oocytes at local level also (Hasan et al., 2014). In aquaculture along with the growth and maturation of gametes, supply of good quality healthy seeds is a crucial requirement as poor quality gametes decrease the rate of fertilization as well as the survival of the progeny (Gallo et al., 2020). However, generation of oxidative cellular stress by the formation of free radicals during gametogenesis resulted deleterious effects on the quality of gametes, is one of the major constrains in fish reproduction. Melatonin, as an antioxidant, is playing a role of potent free radical scavenger (Moniruzzaman et al., 2016; Gozdowska et al., 2022), can minimize the cellular stress generated impairment in both the gonads and ultimately improve the quality of male and female gametes (Mondal et al., 2017; Acharyya et al., 2021; Félix et al., 2021).

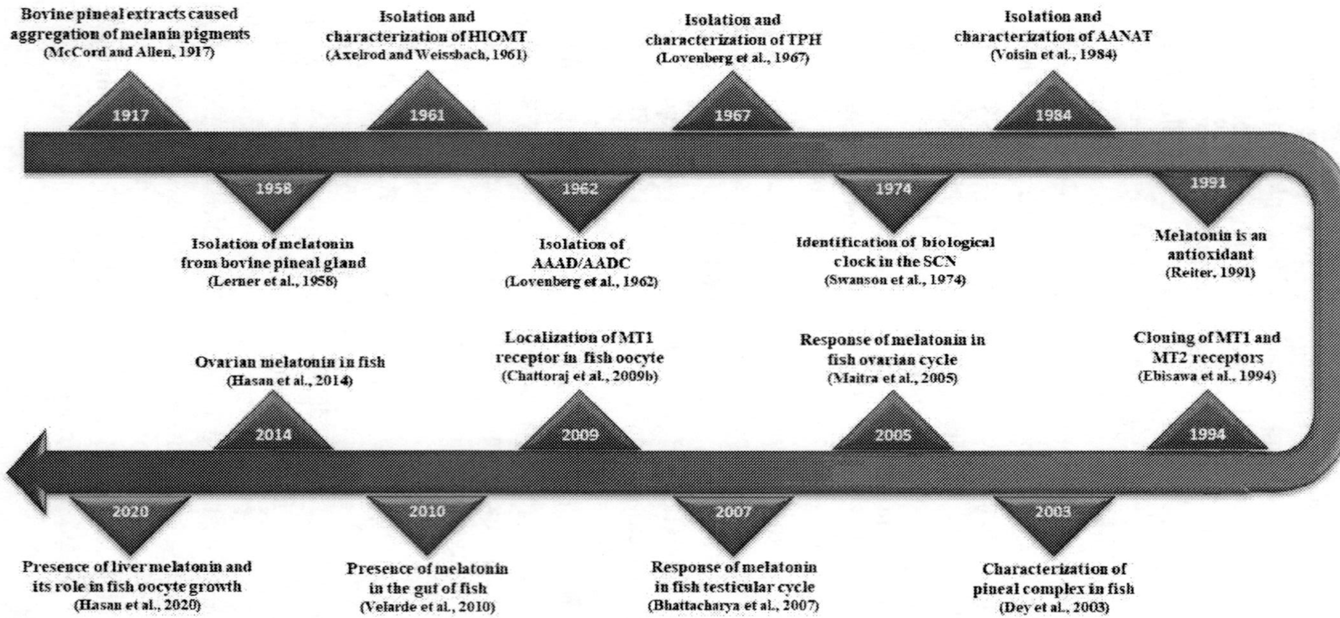

Figure 1. Flow diagram demonstrating the milestones of melatonin: from its discovery and biosynthesis to its pleiotropic effects on seasonal reproduction with special emphasis in fish.

In the state of current knowledge, this chapter is thus an attempt to review the available recent data and observations obtained from natural as well as different specific experimental conditions to provide a comprehensive information and hypothesis regarding mechanism of action of melatonin in gonads and its significant implications in the regulation of fish reproduction in aquaculture.

2. Biosynthetic and Catabolic Pathway of Melatonin

In all the studied vertebrates, melatonin is synthesized from aromatic amino acid L-tryptophan (L-Trp) which is taken up primarily by the pinealocytes of the pineal gland from the circulation. The biosynthesis pathway of melatonin is an enzyme catalyzed four steps reactions. It begins with the hydroxylation of tryptophan in the mitochondria by '*tryptophan-5-monooxygenase/ hydroxylase*' (TPH; EC 1.14.16.4) and is converted into 5-hydroxytryptophan (5-HTP). It is then decarboxylated in the cytosol by '*aromatic amino acid decarboxylase*' (AAAD/AADC; EC 4.1.1.28) into 5-hydroxytryptamine (5HT) or serotonin. Serotonin can act as a substrate for three separate biosynthetic pathways: (1) 5-HT or serotonin can be deaminated by the enzyme '*monoamine oxidase*' (MAO; EC 1.4.3.4) to produce 5-hydroxyindole-3-acetaldehyde (5-HIAL). This indole compound is then either oxidized by an '*aldehyde dehydrogenase*' (EC 1.2.1.3) to form 5-hydroxyindole acetic acid (5-HIAA) or reduced by an alcohol dehydrogenase (EC 1.1.1.1) into 5-hydroxytryptophol (5-HL). Both 5-HIAA and 5-HL are successively *O*-methylated by HIOMT to form 5-methoxyindole acetic acid (5-MIAA) and 5-methoxytryptophol (5-ML), respectively (Falcón et al., 2011). (2) Serotonin can be *O*-methylated directly by '*hydroxyindole-O-methyltransferase*' (HIOMT; EC 2.1.1.4) to produce 5-methoxytryptamine (Axelrod and Weissbach, 1961). (3) The most physiologically important fate of serotonin is the formation of melatonin in two-step reaction. First, serotonin is *N*-acetylated by '*arylalkylamine N-acetyl transferase/serotonin N-acetyltransferase*' (AANAT; EC 2.3.1.87) to form *N*-acetyl serotonin (NAS). AANAT plays a critical role in melatonin biosynthesis as this key enzyme is considered as the "rate limiting enzyme" in this biosynthesis pathway (Klein, 2007). At the final step, NAS is *O*-methylated by HIOMT which leads to the formation of melatonin (*N*-acetyl-5-methoxytryptamine) (Axelrod and Weissbach, 1961) (Figure 2).

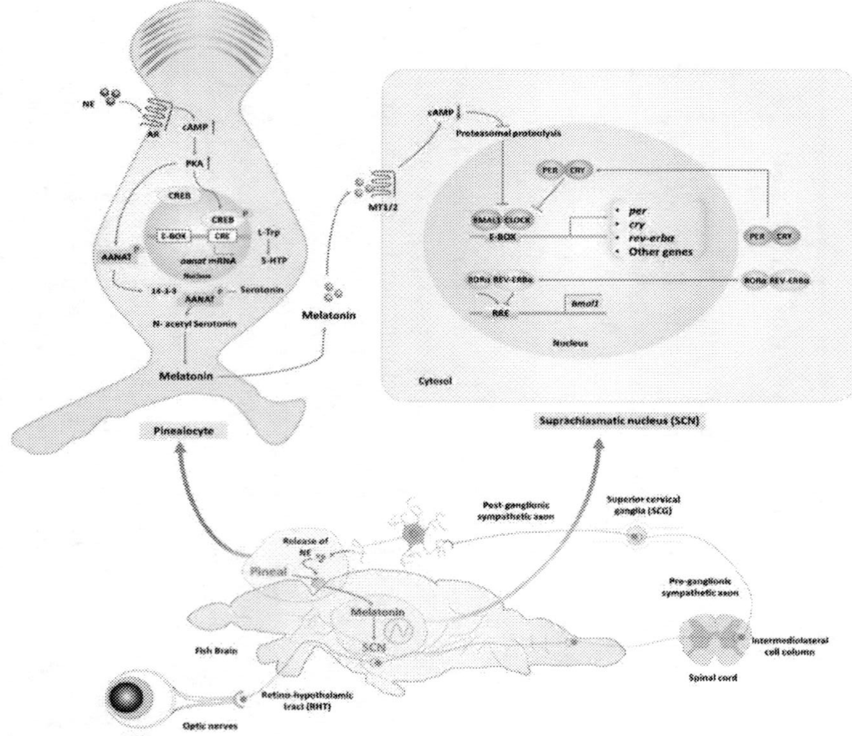

Abbreviations: AANAT: Arylalkylamine-*N*-acetyltransferase; cAMP: cyclic AMP; AR: Adrenergic receptor; BMAL1: Brain and muscle aryl hydrocarbon receptor nuclear translocator-like protein 1; CLOCK: Circadian locomotor output cycle kaput; CRE: cAMP response element; CREB: cAMP-response element binding protein; CRY: Cryptochrome; 5-HTP: 5-Hydroxytryptophan; MT1/2: Melatonin receptor 1/2; NE: Norepinephrine; PER: Period; PKA: Protein kinase A; REV-ERBα: Nuclear receptors reverse erythroblastosis virus-alpha; RHT: Retino-hypothalamic tract; RORα: RAR-related orphan receptor alpha; RRE: REV-ERBα/ROR response element; SCG: Superior cervical ganglia; SCN: Suprachiasmatic nucleus; L-Trp: L-Tryptophan.

Figure 2. The harmony: photoperiod-melatonin-clock genes. (A) Light-dark signal passes through the RHT to SCG. During dark phase, post-ganglionic pineal innervations stimulate the NE mediated melatonin synthesis. (B) Melatonin biosynthesis pathway within pineal organ. NE induces α and β AR mediated cAMP formation in the pinealocyte for further activation of PKA. PKA induces phospho-CREB mediated AANAT synthesis and on other side it phosphorylates AANAT protein to increase its affinity for the binding with 14-3-3 regulatory protein and the formation of a stable AANAT/14-3-3 complex stimulates subsequent melatonin synthesis. (C) In the SCN, rhythmic clock gene expression is intricately regulated by transcription-translation feedback loop consisting of several clock genes and nocturnal expression of such genes is initiated by the BMAL1/CLOCK heterodimer. This expression is halted during day-time by PER/CRY heterodimer which inhibits BMAL1/CLOCK. Proteasomal proteolysis is another vital factor in limiting the expression by degrading BMAL1 and CLOCK proteins. Melatonin through its receptor dependent pathway, decreases intracellular cAMP, and inhibits the proteasomal proteolysis hence stabilize the nocturnal BMAL1 and CLOCK proteins. (*Upward arrow indicates increase and downward arrow indicates decrease*).

The rhythm of melatonin synthesis within pinealocytes is strictly regulated by the degree of expression as well as interaction of circadian clock genes located in the suprachiasmatic nucleus (SCN), circadian oscillator/master pacemaker, of the hypothalamus in brain (Pal et al., 2012). In fish pineal organ, the regulation of melatonin synthesis is usually controlled through the eye and de-pigmented pineal window in response to environmental light-dark signal. The nocturnal increase of nor-adrenaline or nor-epinephrine secreted by the post-synaptic neuronal projections from superior cervical ganglia (SCG) stimulates the α_1- and β_1- adrenergic receptors in pinealocytes. The stimulation of α_1 and β_1-adreno receptors results into an increase of intracellular calcium ion (Ca^{2+}) concentration and adenylate cyclase (AC) enzyme activity, respectively (Chowdhury and Maitra, 2012). The up-regulation of intracellular cAMP leads to a rapid increase of active protein kinase A (PKA), which phosphorylates and protects AANAT protein by the formation of pAANAT/14-3-3 (regulatory protein) complex. Exposure to light during the day, low cAMP level leads to the dephosphorylation of AANAT and subsequent breakdown of the AANAT/14-3-3 complex, results in decrease of the catalytic activity of AANAT by proteasomal degradation of the enzyme (Zhang et al., 2021) (Figure 2).

The mechanism of melatonin catabolism is less understood and the conjugation steps are identified in mammals. In brief, melatonin from blood circulation is primarily metabolized within liver by the enzymes '*cytochrome P450 monooxygenases*', mainly CYP1A2 and converted into 6-hydroxy-melatonin which may conjugate with sulphate to form 6-sulfatoxy-melatonin by '*sulphotransferase*' ST1A3 and finally eliminated through the urine, or it may undergo conjugation with glucuronic acid to form 6-hydroxy-melatonin glucuronide. Urinary measurement of 6-sufatoxymelatonin can be used to determine the pineal melatonin production. In the central nervous system (CNS), melatonin is catabolized to form *N1*-acetyl-*N2*-formyl-5-methoxykynuramine (AFMK) via oxidative reactions by the enzymes '*indoleamine-2, 3-dioxygenase*' (IDO) and/or '*myeloperoxidase*' (MPO). AFMK acts as the primary active metabolite of melatonin. AFMK is further deformylated to form *N*-acetyl-5-methoxy-kynuramine (AMK) by '*arylamine formamidase*' (Amaral and Cipolla-Neto, 2018).

3. Sources and Distribution of Melatonin

3.1. The Pineal: Primary Endogenous Source of Melatonin

The pineal organ or epiphysis cerebri is considered as a self-sustaining 'clock' which plays vital roles in regulating the biological rhythmic functions (Maitra et al., 2013). It is a very small endocrine gland which consists of masses of both sensory neuroglia and secretory endocrine cells called pinealocytes (Vollrath, 1981). The pineal gland in higher vertebrates present dorsally in the epithalamic area of the brain and acts as a typical endocrine gland. However, in lower groups of vertebrates, especially in fish, this organ is present within a window below the skull, through which light can easily enter and stimulates cone-like photoreceptor cells which has a structural analogy with the retinal cones (Gupta, 2016; Saha et al., 2019) and is known to perform both sensory and secretory functions (Dey et al., 2003; Falcón et al., 2010; Maitra et al., 2015). Photoperiodic stimuli in fish also perceived through eye and transmitted through the retino-hypothalamic tract (RHT), via a complex pathway involving suprachaismatic nuclei in hypothalamus to the pineal organ (Maitra and Hasan, 2016). Fish pineal appears to be complex structure made up of several components which lies both within and outside the brain, as compared to higher vertebrates in which the pineal is a discrete organ in the brain (Bhattacharya et al., 2003). Piscine pineal complex exhibits a wide range of anatomical and structural variations and is generally composed of three distinct interconnected parts: an end vescicle (EV), a pineal stalk (PS) and a dorsal sac (DS) (Bhattacharya et al., 2003). In teleost *Catla catla*, DS is highly folded plexiform saccular structure; EV is an antero-laterally elongated, dorso-ventrally flattened vesicular body and connected to the brain by a very long PS (Bhattacharya et al., 2003). However, immunocytochemical localization of melatonin secretory cells suggested their presence at only in the EV of the pineal in Catla (Maitra et al., 2005).

In melatonin biosynthetic pathway, photosensitive protein arylalkylamine-*N*-acetyltransferase (AANAT) is the rhythm generating enzyme (Klein, 2007; Falcón et al., 2014) that determines the cyclical synthesis of melatonin in most of the fish pineal organ (like, goldfish, carp, pike, trout, and white sucker). Photoreceptive fish pineal organ after receiving the light information from the external environment conveys the message by modulating the amount of melatonin synthesis within the pinealocytes, and also acts to co-ordinate the activity of the animals with the external environmental conditions, irrespective of the habit and habitat of the species (Maitra et al., 2015). In fish, *aanat2* gene is expressed specifically in the pineal gland, moreover, *aanat1a* and *aanat1b* has also been detected in dorsal sac of the pineal organ (Isorna et al., 2011; Paulin et al., 2015). However, the temporal organizations of AANAT synthesis in the pineal organ through the circadian axis in different teleosts are not identical (Bégay et al. 1998). In esocidae (pike, *Esox lucius*), cyprinidae (zebrafish, *Danio rerio*), and sparidae (seabream, *Sparus aurata*) group of fish rhythmic expression of AANAT mRNA is regulated by the endogenous clock in the pineal but in salmonids (e.g., rainbow trout, *Oncorhynchus mykiss*) it is independent of any clock-driven changes (Bégay et al., 1998; Iigo et al., 2007a, b). In carp *Catla catla*, 23 kDa AANAT2 protein is present in pineal and its amplitude is highest during dark phase irrespective of reproductive season in an annual gonadal cycle (Figure 3; Table 1). This temporal pattern of AANAT in carp pineal is positively correlated with the diurnal variations of melatonin (Seth and Maitra, 2010). In pineal, diurnal peak concentration is observed at 06:00 h for preparatory and at 24:00 h for the remaining reproductive phases and maximum seasonal value is found in post-spawning phase at 24:00 h (Seth and Maitra, 2010).

3.2. Extra Pineal Sources of Melatonin

3.2.1. Retina

The retina is considered as one of the major extra-pineal sites in vertebrates where endogenous melatonin synthesis takes place. In fish, melatonin secretory cells are localized by immunocytochemical technique in the outer nuclear layer (ONL), outer plexiform layer (OPL) and inner nuclear layer (INL) of the retina (Falcón and Collin, 1991). Two different types of cells are involved in melatonin production: photoreceptor cells and Müller cells. Immunofluorescence study also indicated that perikarya of the ONL, the synaptic processes of photoreceptor cells of the OPL are rich in melatonin content (Falcón and Collin, 1991). In most of the vertebrate species so far investigated, retinal melatonin is produced maximally during darkness and lowest during light phase as it occurs in the pineal organ in a 24 h light-dark cycle. Diurnal rhythmicity of retinal melatonin was first reported in 2-day old rat pups by using specific melatonin antisera (Bubenik et al., 1978). Several studies thereafter also reported on the production of melatonin and its rhythmic pattern in retina of different animal species (Tosini and Menakar

1998). Endogenous rhythmic synthesis of melatonin in retinal photoreceptor cells are primarily regulated by the presence and activity of AANAT, the penultimate enzyme in the melatonin biosynthesis pathway (Chowdhury and Maitra, 2012; Acuña-Castroviejo et al., 2014). Like pineal organ, retinal AANAT activity and melatonin concentrations are strictly controlled by the environmental light-dark driven endogenous circadian clock (Iuvone et al., 2005; Falcón et al., 2010; Chowdhury and Maitra, 2012). AANAT1 is generally responsible for the rhythmic secretion of melatonin in the retina like AANAT2 in the pineal organ (Falcón et al., 2003; Paulin et al., 2015). Among vertebrates, teleost fish has a special characteristic that one or two different varieties of *aanat1* genes may be expressed namely *aanat1a* and *aanat1b* in the retina (Falcón et al., 2003; Isorna et al., 2011; Saha et al., 2019). The nocturnal peak of melatonin level is not exclusive for all animal species, in some animals the rhythm is observed at different time points of the environmental light-dark cycle including the day or no rhythm at all (Iigo et al., 1997; Besseau et al., 2006; Falcón et al., 2011). In zebrafish (*Danio rerio*), goldfish (*Carassius auratus*), carp (*Catla catla*) nocturnal pattern of retinal melatonin production is observed (Iigo et al., 1997; Maitra et al., 2015) but in sea bass (*Dicentrarchus labrax*) retinal melatonin rhythm is changed throughout the seasons (García-Allegue et al., 2001). In carp, highest melatonin concentration in retina is found at mid-night throughout an annual cycle and seasonal peak concentration is observed in post-spawning phase (Maitra et al., 2015). The differences observed in the rhythm of retinal melatonin may be due to the involvement of other retinal cell types like photoreceptors, inter-neurons and ganglionic cells where melatonin biosynthesizing enzymes AANAT and HIOMT are expressed (Besseau et al., 2006; Vuilleumier et al., 2007). Studies in mammals have also provided evidence that the rhythm of melatonin secretion in the retina is altered by the manipulation of photoperiodic conditions such as in continuous light (LL) and continuous darkness (DD) (Tosini and Menaker, 1996) but such type of alteration is varied in fish as non-visual retinal photoreceptors are detected in the inner nuclear and ganglion cell layers (Foster and Bellingham, 2004; Bellingham et al., 2006). Thus, the above-mentioned observations suggest that melatonin synthesis in the retina is under the control of circadian clock as in pineal organ but it may not be the general rule for all species (Iuvone et al., 2005; Falcón et al., 2010). Melatonin is believed to be involved in the regulation of various important functions through autocrine, paracrine or

Table 1. Temporal pattern of endogenous melatonin in the pineal and extra-pineal organs (retina, gastro-intestinal tract, liver, and ovary) and their major possible functions in an annual reproductive cycle of carp *Catla catla* (*Prep: Preparatory phase, Pre-sp: Pre-spawning phase, Spawn: Spawning phase, Post-sp: Post spawning phase*)

Endogenous sources of melatonin	**Average range of melatonin (considering diurnal peak value in each respective phase)**				**Diurnal rhythmicity**	**Seasonal rhythmicity**	**Possible synchronizer(s)**	**Possible effect(s)**	**Reference(s)**
	Prep	**Pre-sp**	**Spawn**	**Post-sp**					
Pineal Organ	40–45 pg/organ	~30 pg/organ	~25 pg/organ	55–60 pg/organ	Peak is observed at 06:00 h during preparatory phase and at 24:00 h during remaining phases of an annual ovarian cycle.	Peak value is found during the post-spawning phase and lowest value during spawning phase.	Environmental light-dark cycle.	Regulates biological rhythm in synchronization with the environmental light-dark cycle.	Seth and Maitra, 2010
Retina	530–560 pg/g tissue	500–525 pg/g tissue	480–500 pg/g tissue	580–610 pg/g tissue	Highest value is observed at 24:00 h irrespective of reproductive phases in an annual cycle.	Peak concentration is observed during the post-spawning phase and lowest value during spawning phase.	Environmental light-dark cycle.	Regulation of retinal circadian clock.	Maitra et al., 2015
Gastro-intestinal tract (GIT)	~450 pg/g tissue	~450 pg/g tissue	~450 pg/g tissue	~450 pg/g tissue	Peak value is observed at mid-day (12:00 h) irrespective of reproductive phases in an annual cycle.	No significant seasonal variations are observed.	Composition and availability of food, feeding time.	Regulates the activity of different digestive enzymes, protects gastrointestinal physiology by reducing oxidative stress against microbial infection etc.	Mukherjee et al., 2014; Pal et al. 2016a, b; Acharyya et al., 2021

Endogenous sources of melatonin	**Average range of melatonin (considering diurnal peak value in each respective phase)**				**Diurnal rhythmicity**	**Seasonal rhythmicity**	**Possible synchronizer(s)**	**Possible effect(s)**	**Reference(s)**
	Prep	**Pre-sp**	**Spawn**	**Post-sp**					
Liver	260–300 pg/g tissue	160–200 pg/g tissue	115–130 pg/g tissue	700–750 pg/g tissue	Diurnal peak is observed at the onset of darkness (18:00 h) irrespective of reproductive phases in an annual cycle.	Seasonal peak value is found during the post-spawning phase and lowest during the spawning phase.	Environmental light-dark cycle.	Stimulates different antioxidative enzymes, regulates reproduction by modulating the levels of hepatic 17β-estradiol and vitellogenin contents.	Hasan et al., 2016; 2020
Ovary	60–70 pg/g tissue	160–175 pg/g tissue	210–250 pg/g tissue	38–45 pg/g tissue	No significant diurnal variations are observed.	Highest intra-ovarian melatonin is observed during the spawning phase and lowest during post-spawning phase.	Reproductive status of the fish.	Antioxidative function which includes scavenger of free radicals, activation of different antioxidative enzymes, promotes oocyte maturation.	Chattoraj et al., 2005; Hasan et al., 2014; Moniruzzaman et al., 2016; Mondal et al., 2017

neuro-endocrine mode in this photo-sensitive retinal organ (Blasiak et al., 2016). The photic signal transduction mechanism shares common components to execute its effects by interacting with G-protein coupled family of receptor proteins (Tosini et al., 2012) (Figure 3; Table 1).

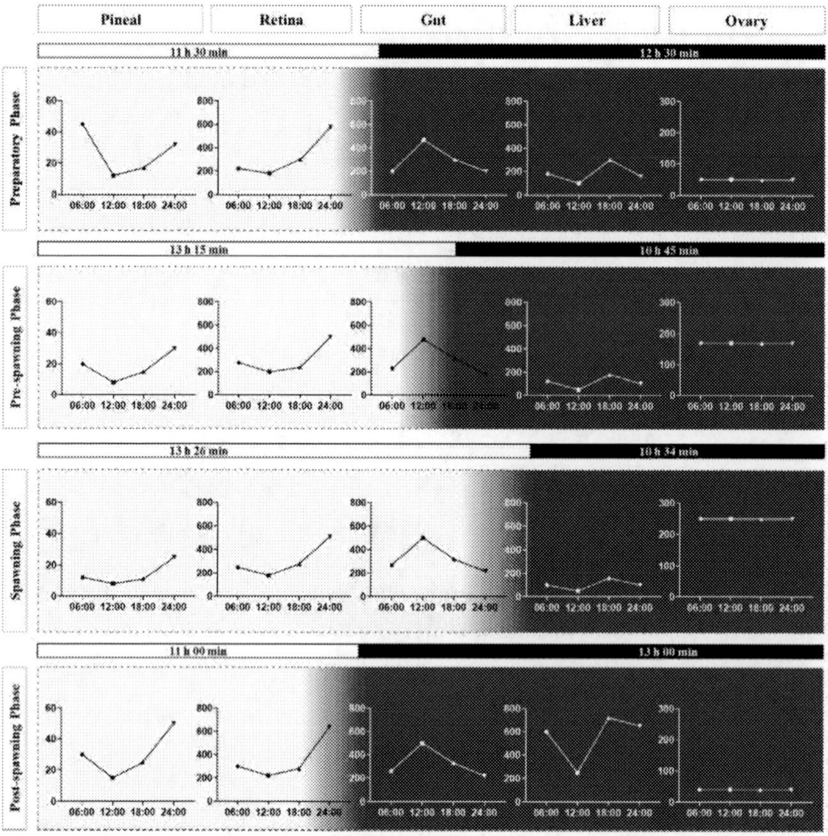

Figure 3. Diurnal and seasonal patterns of melatonin in pineal and various extra-pineal organs in fish. Graphical representation of the diurnal profiles of melatonin concentrations in the pineal, retina, gut, liver and ovary at four different time points (06:00, 12:00, 18:00, and 24:00 h) in a 24 h cycle during four different reproductive phases (preparatory, pre-spawning, spawning and post-spawning) in an annual cycle in carp *Catla catla*. At the top of all graphical illustrations, open and solid horizontal bars represent the period of light and darkness in each phase, respectively.

3.2.2. Gastro-Intestinal Tract (GIT)

In vertebrates, the gut or gastro-intestinal tract (GIT) is also considered as another major extra-pineal site of melatonin production (Bubenik, 2002). With the use of specific antibody against melatonin and highly sensitive immunocytochemical techniques, many researchers exhibited cellular localization, distribution as well as functions of gastrointestinal melatonin largely in mammalian and avian species. Melatonin in gut was first reported in rats (Raikhlin and Kvetnoy, 1974) and thereafter subsequent studies revealed the presence of melatonin producing cells in the GIT of several groups of vertebrates starting from mammals (Bubenik and Brown; 1997), birds (Vakkuri et al., 1985), to reptiles, amphibians and fish (Bubenik and Pang, 1997; Velarde et al., 2010). Earlier, it was proposed that enterochromaffin cells of mammalian gastrointestinal tract (GIT) may synthesize or produce melatonin (Raikhlin et al., 1976). Further studies confirmed the gene expressions of key melatonin biosynthesizing enzymes, AANAT and HIOMT, and suggested that the production of this enteric hormone is independent of the pineal gland (Velarde et al., 2010; Devi et al., 2016). GIT melatonin concentration is 10-100 times higher than in the serum (Bubenik et al., 1996) and is 400 times higher than the pineal gland (Huether, 1993). Although the concentrations may vary in between different parts of the gut in rat and in mouse (Bubenik et al., 1992) but such type of segmental variations are absent in carp *Catla catla* (Mukherjee et al., 2014a). Studies on the same carp, the levels/abundance of melatonin and 23kDa AANAT protein varies diurnally with a peak during the midday (12:00 h) under natural as well as artificial photoperiodic conditions (Mukherjee et al., 2014 a,b) (Figure 3; Table 1). Congruently, peak melatonin titre in gut is found at 12:00 h throughout an annual cycle, though it is seasonally arrhythmic (Mukherjee et al., 2014 a,b). From this observation, it seems likely that cyclic gut melatonin synthesis is under the regulation of non-photic environmental cues, like availability and composition of food, feeding time (Konturek et al., 2011; Acharyya et al., 2021). The functions of GIT melatonin are not fully understood yet but it seems different than that of the pineal melatonin which acts as a hormone while GIT melatonin may act locally through an autocrine or paracrine manner (Maitra et al., 2015; Pal and Maitra, 2018). A very few recent reports demonstrate that GIT melatonin stimulates different digestive enzymes (amylase, protease, lipase) and helps in the digestive process in carp (Pal et al., 2016a; Pal and Maitra, 2018). GIT melatonin has significant role in the reduction of tissue damage during microbial infection in fish by stimulating various antioxidative enzymes (Pal et al., 2016b), so the practice of this gut derived hormone in aquaculture may not be ruled out (Acharyya et al., 2021).

3.2.3. Liver

Liver is the most vital peripheral organ where circulating melatonin (92–97%) is catabolised. However, it was believed that melatonin comes from blood is the only source of liver melatonin but whether or not the hepatic tissue is the endogenous source, that was unresolved. In 1993, by using immunohistochemical technique, Menendez-Pelaez and his colleagues detected melatonin for the first time in the nuclei of hepatocytes of mammals (Menendez-Pelaez et al., 1993). Hepatic melatonin concentration is 15 times higher than the blood circulation and most of it is derived from GIT (Bubenik, 2002). Later, a very few studies reported that the key enzyme, AANAT, of melatonin biosynthesis pathway is present in mammalian and non-mammalian hepatocytes including fish (Nowak et al., 1995; Velarde et al., 2010; Hasan et al., 2016). So, melatonin is considered to be synthesized endogenously in liver (Maitra et al., 2015). In carp *Catla catla*, both hepatic AANAT protein (23 kDa) and melatonin levels show distinct parallel diurnal and seasonal rhythmicity with a peak at the onset of darkness (18:00 h) and nadir at midday in each phase of an annual reproductive cycle under natural photo-thermal conditions and seasonal peak in melatonin concentration is found in post-spawning phase (Hasan et al., 2016) (Figure 3). The result significantly depicted that hepatic melatonin synthesis may be synchronized with the photoperiodic conditions (Table 1). The first ever study by the same researchers also demonstrated that liver melatonin regulates ovarian functions by controlling two major axes in any fish. It controls 'hormonal axis' by interacting with two important indices of oocyte growth, sex steroid 17β-estradiol and vitellogenin protein (~115 and 102 kDa), and 'stress axis' by activating different enzymatic and non-enzymatic antioxidants for the reduction of oxidative free radical load (Hasan et al., 2016, 2020).

3.2.4. Male Gonad or Testis

To our knowledge, information regarding the melatoninergic system in male gonad or testis is scarce. No such data is available about the source and distributions of melatonin in testis but a few numbers of experimental evidences support that melatonin may be synthesized endogenously within male reproductive organ as *aanat* gene expression is detected in human testis (Albrecht, 2009; Huang et al., 2022). However, a very recent study on fish mRNA expression and tissue-specific immunohistochemical localization of

AANAT and HIOMT, endogenous testicular melatonin synthesizing parameters, along with serum and testicular melatonin concentrations have well documented in case of zebrafish (*Danio rerio*) in a 24 h light-dark cycle (Devi et al., 2021). Interestingly, testicular melatonin has a day-night variations in concentrations and the observation collectively revealed that there is a possible involvement of melatonin to modulate the testicular physiology in an autocrine and/or paracrine manner (Acharyya et al., 2021; Devi et al., 2021). However, ample data on the melatoninergic system under varied experimental conditions are required for an unequivocal demonstration of the role of environmental cues as a possible synchronizer of melatonin rhythms in the testis of any fish species.

3.2.5. Female Gonad or Ovary

Female reproductive organ, ovary, is an additional important extra-pineal source of melatonin. Although melatonin is detected and measured in the follicular fluid (FF) for quite a long time in mammals (Brzezinski et al., 1987; Ronnberg et al., 1990) but the question is whether melatonin is synthesized within the ovary itself or, is taken up from the circulation remained enigmatic. On the other hand, blood circulation may supply a large amount of melatonin measured in the ovary and FF, as the rat and cat ovaries have shown the ability to pick up and retain circulating [^{3}H]-melatonin (Tamura et al., 2020). Interestingly, studies have also found that melatonin concentration in human FF exceeds the amount of circulatory melatonin and the result may be a sign of the incorporation of circulatory melatonin in the ovary (Reiter et al., 2013). Presence of melatonin and its immediate precursor *N*-acetylserotonin (NAS) in the ovarian extracts of humans indicates the possibility of an endogenous melatonin synthesizing system in the ovarian FF (Itoh et al., 1999). Melatonin biosynthetic enzyme 'acetylserotonin O-methyltransferase' (ASMT) is identified in the cumulus-oocyte-complex (COC) of bovine follicles (Salhab et al., 2013) provides another important indication that ovary may synthesize melatonin which has an important role in the regulation of ovarian functions such as oocyte maturation, ovulation, steroidogenesis, etc. However, due to the lack of comparable data from any prior study, knowledge on ovarian melatonin in any fish species remained poignant. In carp *Catla catla* intra-ovarian melatonin concentration is measured for the first time, but the use of specific antibody against AANAT protein in Western blot do not show any band and thereby raised question on the existence of melatonin synthesizing machinery within the ovarian follicles (Hasan et al., 2014) (Table 1). On the contrary, a recent study also reported that genes of melatonin biosynthesizing enzymes, *Tph1a, Aanat1, Aanat2 and Hiomt* are expressed in the ovary of zebrafish (Khan et al., 2016). Moreover, temporal pattern study of ovarian melatonin demonstrated that it has no significant daily variations in natural habitat regardless of fluctuations in environmental light and temperature experienced by carp. The result thereby rules out possible role of environmental photo-thermal conditions as a synchronizer for ovarian melatonin synthesis. The study also explained seasonal fluctuations of ovarian melatonin titer that during spawning it is approximately five times higher than measured during post-spawning phase (Figure 3). This observation promotes the idea of a possible role of developmental as well as endocrine status of the ovary in determining its seasonal melatonin profiles in carp. The findings are consistent with the report on the FF of human ovary, melatonin levels are two times higher in large follicles just before ovulation compared with those in smaller antral, immature follicles (Nakamura et al., 2003), so melatonin may have an important function during the maturation of ovarian follicles and ovulation.

4. Rhythmicity of Melatonin in an Annual Reproductive Cycle of Fish

The annual reproductive cycle of seasonally breeding fish comprises of a long preparatory phase and a short spawning phase when the gonads reach to their maturity. On the basis of germ cell profiles, the gonadal cycle is divided into four distinct phases in carp *Catla catla* (as a seasonal breeder): (a) the preparatory phase (January-March), (b) pre-spawning phase (April-June), (c) spawning phase (July-August), and (d) post-spawning phase (September-December). Each reproductive phase has a specific duration and is characterized by distinct germ cell profiles in the ovary (Dey et al., 2004) and in the testis (Bhattacharyya and Maitra, 2006). In an ovarian cycle, the preparatory phase with mostly oogonia and few stage-I primary oocytes is followed by the pre-spawning phase identified by the presence of abundant number of stage-II oocytes as the most advanced germ cells. The spawning phase exhibits yolk containing mature stage-III and stage-IV oocytes but the post-spawning phase contains a large number of atretic or post-ovulatory follicles (Hasan et al., 2014). In a testicular cycle of carp, immature stage-I spermatids as the most advanced stage during preparatory phase and the development of germ cell profiles beyond stage-I spermatids occurs during pre-spawning phase. The most mature germ cell profiles in testis are attained during the spawning phase which is identified by the presence of mostly spermatozoa and then a sudden drop in the mature germ cell profiles occurs during the post-spawning phase of the annual cycle (Bhattacharya et al., 2007; Maitra et al., 2015). In most of the fish, periodic changes of different environmental cues such as photoperiod along with water temperature and other physico-chemical components determine the species specific temporal patterns of seasonal reproduction. The photoperiodic signal, acting as a most vital critical cue of biological rhythm, is processed by the neuroendocrine system to synchronize with different circannual rhythmic functions associated with seasonal reproduction in fish (Dey et al., 2004). Melatonin, after its synthesis from the photosensitive pineal organ, is released into the blood circulation and reaches to various target organs, including both male and female gonads, where it acts through specific cellular mechanisms (Chattoraj et al., 2005). From the observations of various sets of experiments, researchers suggested that the pattern of diurnal rhythmicity of melatonin is varied according to the developmental status of gonads in fish (Maitra et al., 2005; Hasan et al., 2016). A large number of studies on lower vertebrates including fish, three different types of nocturnal melatonin rhythms are observed, namely Type-A, -B and -C (Falcón et al, 2010; Migaud et al., 2010). The Type-A rhythm is identified by a discrete peak of melatonin concentration in the late-dark phase as seen in Atlantic cod *Gadus morhua* (Porter et al., 2000) and Haddock *Melanogrammus aeglefinus* (Davie et al., 2007). The Type-B rhythm is characterized by a discrete peak in the mid-dark phase as seen in Nile tilapia *Oreochromis niloticus niloticus* (Matinez-Chavez et al., 2008) and C-Type rhythm is characterized by a rapid increase of melatonin in the early-dark phase or after the onset of darkness as seen in Atlantic salmon *Salmo salar*, Rainbow trout *Onchorhynchus mykiss* and Atlantic halibut *Hippoglossus hippoglossus* and other teleosts (Falcón et al., 2010; Migaud et al., 2010; Maitra et al., 2005). The reason behind the different types of nocturnal melatonin profiles observed in different species is not clearly understood. Studies on carp reported for the first time that the nocturnal pattern of serum melatonin profiles vary from Type-A to Type-B in the same species depending upon the reproductive status of the animal (Maitra et al., 2013; Hasan et al., 2014). The Type-A rhythm is confined only during the preparatory phase and Type-B rhythm is observed during rest of the phases in an annual reproductive cycle as observed during pre-spawning, spawning and post-spawning phases (Maitra et al., 2005). This type of rhythmic variations and reproductive stage specific 'phase shift' of the rhythm may be due to the seasonal changes of some specific environmental parameters like duration and intensity of photic signal, temperature and salinity (Maitra et al., 2015). The rhythmicity of melatonin production also depends greatly upon the self-sustained endogenous clock machinery within the organ.

5. Role of Melatonin Regulated 'Clock Genes' in Rhythmic Functions

Melatonin acts as an important regulator of circadian rhythms by acting as a "zeitgeber" to nearly all the tissues and organs. The circadian clock functioning through the formation of a loop and is controlled by an interlocking network of 'transcription-translation feedback loop' (TTFL) mechanism (Vriend and Reiter, 2015). Generally, the TTFL system has two arms: a positive arm, where the transcription factors promote the transcription of clock-controlled genes (CCGs) and a negative arm in which the products of positive arm act as repressors and negatively regulate their own expression. There are at least ten genes found to be responsible for the regulation of circadian rhythms which include *Clock, Bmal1, Per1, Per2, Per3, Cry1, Cry2, Cry3, Dec1, Dec2* and *Rev-erbα.* These genes possess conserved 'E-Box' element in their promoter region. In vertebrates, two important nuclear transcription factors, 'circadian locomotor output cycle kaput' (CLOCK) and the 'brain and muscle aryl hydrocarbon receptor nuclear translocator-like protein1' (BMAL1) initiate the functions of TTFL. CLOCK and BMAL1 proteins together form a heterodimeric transcriptional activator (Doi et al., 2006), which translocates into the nucleus for binding to specific 'E-box' elements (CACGTG) in the promoter region of CCGs and promotes the transcriptions of the specific genes (Chowdhury and Maitra, 2012) (Figure 2). The clock gene machinery is well established in the regulation of rhythmic functions in mammalian system but such comparable studies are so inadequate in fish species. In mammals, cytoplasmic PER proteins are phosphorylated by a group of casein kinase-I proteins, CKIδ and CKIϵ, which make it stable dimer with CRY protein (Akashi et al., 2002). The CRY protein is considered as the rate limiting repressor of transcription (Kume et al., 1999), which translocates and binds with BMAL1 protein into the nucleus to represses the function of it, thereby completes the circadian loop (Mattam et al., 2014). DEC1 and DEC2 also repress the transcriptional activation of *Per* and *Cry* genes (Kennaway, 2005). According to various researchers, *Bmal1* is the most studied component of the molecular circadian clock among the core clock genes and it is found that dysfunction of BMAL1 affects both cellular and behavioural circadian rhythms in a most dramatic manner jeopardizing the circadian clock (Pan et al., 2020). An additional loop, generated through REV-ERBα and RORα, also controls the expression of clock genes. However, the RORα seems to regulate the expression of *Bmal1* positively, therefore promotes the loop in the forward direction (Akashi and Takumi, 2005). Melatonin may also regulate the circadian clock machinery by acting as a proteasomal inhibitor of clock proteins (Vriend and Reiter, 2014). The molecular mechanisms involved in the central and peripheral clocks are generally same but the components and their interactions of different clock genes may vary in a tissue specific manner.

5.1. Melatonin and Clock Genes in Fish Reproduction

Ovary is considered to be one of the key peripheral organs controlled by the central pacemaker or clock situated within the SCN and is of prime importance for the better understanding of reproductive physiology in a seasonal breeder. Most of the studies indicating that clock genes are expressed in the ovary are limited in mammals (Sellix, 2015; Brzezinski et al., 2021), but a very few recent observations demonstrated that rhythm controlling clock genes are also present in fish ovary (Li et al., 2013; Khan et al., 2016). In zebrafish, all of the clock genes are found to be expressed in the ovary in a rhythmic fashion in various artificial photoperiodic conditions and therefore it is argued that the ovary functioning as a peripheral circadian oscillator which can work in synchronization with the central oscillating system (Li et al., 2013; Khan et al., 2016). The researchers also suggested that the ovary of zebrafish can synthesize endogenous melatonin by its own melatonin biosynthesizing machinery as a result of the synchronization between the central and the peripheral clocks (Khan et al., 2016). Melatonin may also play a role in the regulation and maintenance of ovarian clock and its associated clock genes in a circadian or circannual manner and thereby may regulate some vital ovarian functions such as oogenesis, LH hormone production, ovulation and steroidogenesis (Selix, 2015). To our knowledge, data on 'testicular clock' are only limited in mammals (Feldberg, 2021) and needs further research to fully unravel its role in testicular functions in non-mammalian animal species.

6. Role of Melatonin in Seasonal Reproduction of Fish

One of the first evidences of the association between melatonin and reproduction comes from the experiments in mammals in which it was found that in male hamsters, exposure to daily short photoperiodic conditions results into gonadal atrophy and pinealectomy prevented the inhibition effects on gonad. Thus, it was concluded that there must be a strong relationship exists between photoperiod, pineal activity and reproduction (Hoffman and Reiter, 1965). Several studies in the last few decades have focused on the role of pineal hormone melatonin in the seasonal reproduction of fish (Maitra et al., 2005). The role of melatonin in determining the sexual status, especially in female, is well documented in non-mammalian vertebrates including fish (Falcón et al., 2010; Maitra et al., 2013). Reproduction in most of the fish, like many other vertebrates, is discontinuous consisting of a short spawning phase and a long phase of preparation. The periodic annual changes of physiological events and environmental parameters are strictly often synchronized with each other that results spawning in a specific suitable part of an annual gonadal cycle to ensure maximum reproductive outcome and better survival of the offspring (Foster and Kreitzman, 2005). Seasonal variations in the duration of the solar day or photoperiod have been proved to be the primary and regular variable that alone, or in combination with other environmental component(s), impel the 'driving function' in controlling the sexual periodicity in most fish species that breed at mid- and high latitudes (Bayarri et al., 2009, Falcón et al., 2010, Zohar et al., 2010), and even at low latitudes (Dey et al., 2004; Bhattacharya et al., 2007). In these respects, a large number of studies on diverse groups of fish have contributed largely to provide persuasive evidence suggesting that melatonin indeed play a pivotal role in the regulation of piscine reproduction.

6.1. In Female

6.1.1. In Oocyte Growth and Development

In non-mammalian vertebrates, growth of oocytes is arrested in the first meiotic prophase. At this stage the principal events responsible for the enormous oocyte growth and development of fish, termed as vitellogenesis. In this cascade of physiological processes, hypothalamus in fish brain secretes a decapeptide hormone, 'gonadotropin releasing hormone (GnRH) which carried by the portal circulation and acts on gonadotroph cells in pituitary to release gonadotropic hormones (GTH I and II) (Bhattacharya, 1999). In most of the teleosts, GTH stimulates somatic theca and granulosa cell types in the ovary to release sex steroids, estradiol-17β and 17α, 20β-dihydroxy-4-pregnen-3-one (17α, 20β DHP) (Nagahama, 1994). 17β-estradiol is transported by binding with sex hormone binding globulin (SHBG) in blood circulation to hepatocyte (Suresh et al., 2008) and forms a ligand-receptor complex with its cytoplasmic estrogen receptors (ERα and ERβ) to activate the transcription of vitellogenin gene (*vtg*) in nucleus (Nagahama and Yamashita, 2008). The vitellogenin or yolk protein is then transported through blood and is incorporated in the developing oocytes by 'vitellogenin receptor' to promote its growth and maturation (Jalabert, 2005). In fish, the amplitude of both 17β-estradiol and 17α, 20β DHP or maturation inducing hormone (MIH) in the serum is maximum during the spawning phase and then a sudden fall is observed during the remaining phases of an annual reproductive cycle in natural photoperiodic conditions (Hasan et al., 2016). A recent report on Indian major carp *Catla catla*, displayed that melatonin is negatively correlated with 17β-estradiol, MIH and vitellogenin in serum and liver in an annual gonadal cycle. This observation is a breakthrough in understanding the role of melatonin in the regulation of oogenesis by modulating the levels of these oocyte growth indices (Hasan et al., 2016) (Figure 4). Researchers also focused on the same carp (Catla) and emphasized that seasonally varied ovarian melatonin levels exert its role in the way of oocyte growth and development by reducing the oxidative stress (Hasan et al., 2014). The target site as well as action mechanism of melatonin in fish reproduction is varied according to fish species (Pal et al., 2012). In Atlantic croaker, *Micropogonias undulates*, melatonin influences LH secretion directly at the pituitary level specifically and indirectly at the total brain level (Khan and Thomas, 1996). A study in female three-spined stickleback (*Gasterosteus aculeatus*) indicates that the major part of the photoperiodic effects is mediated through a mechanism other than involved in case of circulating melatonin (Bornestaf et al., 2001).

Exogenous melatonin administration in fish significantly depicted that this indoleamine controls the reproductive seasonality by stimulating the final stages of maturation by synchronizing with the optimal timing of spawning (Bromage et al., 1994). However, the study on the effects of melatonin administration during different phases of reproductive cycle in female carp (*Catla catla*) demonstrated that influences of exogenous melatonin vary in

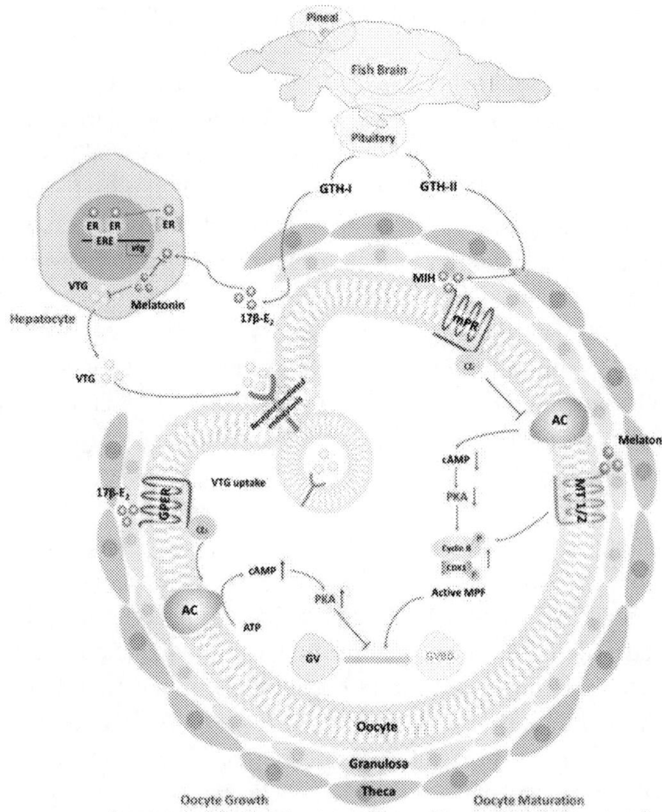

Figure 4. Melatonin mediated regulation of oocyte growth and maturation in fish. Gonadotropin (GTH-I) mediated release of 17β-estradiol from the ovary and its binding with estrogen receptor (ER) exerts hepatic vitellogenin (VTG) release into blood. 17β-estradiol through another signaling pathway inhibits oocyte maturation by coupling with membrane bound G-protein coupled estrogen receptor (GPER) and subsequent increase of adenylate cyclase (AC) mediated intracellular cAMP. The ultimate result helps in oocyte growth. Melatonin negatively regulates the action of 17β-estradiol and synthesis of VTG in liver. Prior to the spawning, GTH-II triggers release of maturation inducing hormone (MIH) which initiates oocyte maturation by governing the 'germinal vesicle breakdown' (GVBD) through non-genomic membrane progestin receptor (mPR). Melatonin augments this maturational process through its membrane bound receptor (MT1/2) in oocytes and accelerates the progression towards GVBD. (*Upward arrow indicates increase and downward arrow indicates decrease*).

Table 2. Effects of exogenous melatonin and artificial photoperiodic conditions on reproduction in different experimental fish species

Experimental regimens	Fish species	Dose(s), Duration and Mode of melatonin exposure/Photoperiodic regimens	Possible Effects	Reference(s)
Exogenous melatonin administration	Female carp, Labeo rohita and Cyprinus carpio	50, 100 or 500 pg/mL in culture media.	Melatonin, 2 h prior, co-administration and 2 h after the addition of MIH increases maturational competence in MIH-induced denuded oocytes.	Chattoraj et al., 2005
	Female carp, Catla catla	100 μg/100 g body weight/day for 15 and 30 consecutive days through intramuscular injection.	Melatonin exerts its pro-gonadal effects during preparatory phase and anti-gonadal during pre-spawning and spawning phases and no response during post-spawning phase in an annual cycle.	Maitra et al., 2005
	Male carp, Catla catla	25, 50, or 100 g/100 g body weight/day for 15 and 30 consecutive days through intramuscular injection.	Intramuscular injection of melatonin results duration dependent pro-gonadal or anti-gonadal effects in an annual testicular cycle.	Bhattacharya et al., 2007
	Female zebrafish, Danio rerio	100 nM and 1mM for 10 days into tank water.	Administration of melatonin stimulates competence and maturation of oocytes as well as the number of ovulated eggs.	Carnevali et al., 2010, 2011
	Female zebrafish, Danio rerio	100 nM and 1mM for 10 days into tank water.	Exogenous melatonin significantly increases the number of vitellogenic and mature ovarian follicles.	Giorgini et al., 2012
	Male catfish, Clarias macrocephalus	50 and 250 mg/kg fish balanced diet into tank water for 8 weeks.	Melatonin administration as balanced fish diet increases the value of GSI, number and viability of spermatozoa indicating precocious maturation of testis.	Aripin et al., 2015
	Female carp, Catla catla	Melatonin (25 μg/100 g body weight) in various combinations with ovaprim through intramuscular injection.	Melatonin pre-treatment (–2 h) of ovaprim increases the rate of oocyte maturation through the formation of MPF and activation of different antioxidative enzymes.	Moniruzzaman et al., 2016

Table 2. (Continued)

Experimental regimens	Fish species	Dose(s), Duration and Mode of melatonin exposure/Photoperiodic regimens	Possible Effects	Reference(s)
Exogenous melatonin administration (continued)	Female carp, Catla catla	100 g/100 g body weight/day for 15 days through intramuscular injection.	Administration of melatonin reduces ROS generation during oocyte growth and maturation by the elevation of different antioxidative enzymes in an annual ovarian cycle.	Mondal et al., 2017
	Male paddlefish, Polyodon spathula	0.1, 0.5, 1, and 5 μM for up to 72 h into sperm storage solution.	In vitro administration of melatonin inhibits the generation of ROS and helps in the maintenance of mitochondrial membrane potential and ATP content in sperm.	Gao et al., 2019
Photoperiodic manipulations	Female catfish, Mystus tengara	Long (14L:10D; 18L:06D) photoperiodic conditions.	Ovarian growth and maturation upregulated under long photoperiodic condition.	Guraya et al., 1976
	Female catfish, Heteropneustes fossilis	Long (LP) and short photoperiods (SP).	LP condition stimulates ovarian growth and the effects are reversed under the influence of SP.	Sundararaj, 1981
	Female murrel, Channa punctatus	Continuous light (24L:00D).	Gonadal growth increases under continuous illuminating condition.	Srivastava and Singh, 1992
	Male murrel, Channa punctatus	Long photoperiod (LP: 16L:08D) and short photoperiod (SP: 08L:16D) for one year.	LP regimen stimulates testicular maturation and the activity of gonadotropic cells, whereas SP regimen inhibits the pro-gonadal action of LP.	Srivastava and Singh, 1993
	Male and female carp, Catla catla	Long photoperiod (16L:08D) and short photoperiod (08L:16D).	Long photoperiod stimulates the steroidogenic properties of the gonad and short photoperiodic condition inhibits the gonadal growth and maturation in an annual reproductive cycle.	Bhattacharyya et al., 2005; Dey et al., 2005
	Female nile tilapia, Oreochromis niloticus	24L:00D, 20L:04D and 18L:06D for 24 weeks	Continuous light decreases GSI and mean oocyte size significantly in compare long photoperiodic conditions and control.	Rad et al., 2006

relation to the dose and duration of hormonal treatment as well as the gonadal status of the fish (Maitra et al., 2005). Information on the effects of exogenous melatonin in adult carp is provided following intra-muscular application of different doses (25-, or 50-, or 100 μg /100 g body weight/day) at the on-set of darkness for 1 day, 15-, and 30 days (Maitra et al., 2005). Exogenous melatonin, irrespective of its dose and duration of treatment, fails to elicit any ovarian response during the post-spawning phase, while no remarkable impact of single administration of melatonin are observed in any part of an annual cycle, though in laboratory rodents a single injection of melatonin is known to exert significant influences on the functions of hypophyseal-gonadal axis (Maitra and Ray, 2000). Each dose of melatonin injected daily for 15 or 30 days during the preparatory phase causes precocious growth of oocytes (formation of stage-II oocytes) compared to the ovaries in control fish showing stage-I oocytes as the most advanced follicles. Conversely, melatonin at each dose causes a significant decrease in the number of stage-II and stage-III oocytes during the pre-spawning phase and spawning phase respectively, with subsequent increase in the number of degenerated or atretic follicles relative to their number in the ovaries of control fish. Thus, it may be concluded that exogenous melatonin administration may result in variable influences, like pro-gonadal or anti-gonadal or no subtle effects on gonads depending on the reproductive status of fish in an annual cycle (Maitra et al., 2005; Mondal et al., 2017). The anti-gonadal effects of melatonin are also known for other tropical fish like *Heteopneustes fossilis* (Sundararaj and Keshavanath, 1976), *Mystus tengara* (Saxena and Anand, 1977), and *Clarias batrachus* (Singh and Lal, 1994), though a pro-gonadal action is also known in additional fish (Singh and Lal, 1994; Bromage et al., 1995). Melatonin administration in *Heteropneustes fossilis* for 20 days after the onset of light phase inhibits vitellogenesis, induces follicular atresia and lowers the number of pituitary gonadotrops during the pre-spawning phase and induces ovarian regression and reduces the number of pituitary gonadotrophs in the spawning phase (Sundararaj and Keshavanath, 1976). In *Mystus tengara* melatonin injection 2 h after the onset of light phase during the phase of ovarian recrudescence arrests vitellogenesis and increases the frequency of atretic follicle (Saxena and Anand, 1977). The study on *Clarias batrachus* reveals that melatonin administration at lower doses of 25- and 50- μg/fish significantly decreases the 17α-hydroxyprogesterone levels and at higher doses (100-, 200- and 400 mg/fish) further reduces the hormone to non-detectable levels (Singh and Lal, 1994). Implantation of melatonin in *Fundulus similes* induces spawning in an asynchronous pattern (Bromage et al., 1995), while the fecundity in zebrafish females is affected positively by exogenous melatonin administration (Carnevali et al., 2010) (Table 2). This is likely that melatonin action interacts with the brain-pituitary-gonadal axis (Amano et al., 2000), or with one or more of a variety of peripheral and/or central sites in fish.

6.1.2. In Oocyte Maturation

Fish oocyte maturation is a naturally occurring event with two successive meiotic divisions in which two polar bodies are released at the end of the cycle (Jalabert, 2005). During first meiotic division, the nucleus of the arrested primary oocyte at the diplotene stage moves towards the peripheral region of the cell followed by the fusion of nuclear and cell membranes and the first polar body is released that ultimately marks the completion of the meiosis-I (Nagahama, 1994). At prophase to metaphase transition, breakdown of oocyte nuclear membrane or germinal vesicle breakdown (GVBD) takes place which is generally considered as the penultimate hallmark of the oocyte maturation in fish (Nagahama and Yamashita, 2008; Lubzens et al., 2010). The second meiotic division, starts immediately after the completion of the first meiotic division, progresses only up to the metaphase stage and resumption of meiosis or final oocyte maturation (FOM) is generally induced by pre-ovulatory gonadotropin surge. During this stage animal and vegetal poles of the oocyte are differentiated clearly. The promoting effects of gonadotropin on oocyte of most the teleosts involve two major mechanisms in two temporally different developmental stages (Patiño et al., 2001): (a) a steroid-independent priming stage that increases the maturational competence of the oocytes, and (b) a maturational stage that depends on the gonadotropin stimulated synthesis of maturation-inducing steroid (MIS) or maturation-inducing hormone (MIH), involving steroidogenic pathways in the ovarian follicular cells (Patiño and Thomas, 1990). In most fish, FOM is initiated by the potent MIH (17α, 20β-DHP) which acts on oocyte membrane receptors and leads to the formation of maturation promoting factor (MPF) in the oocyte cytoplasm (Nagahama, 1997). The molecular structure of MPF shares similarity among different vertebrates which is composed of two subunits- (i) cyclin B, a regulatory subunit and (ii) cyclin-dependent kinase (Cdk1, or Cdc2, or p34 kinase), a catalytic subunit of which the activation of cdc2 is related with its mobility shift from 35 to 34 kDa (Chattoraj et al., 2005; Nagahama and Yamashita, 2008). Study performed in goldfish on the formation and activation of MPF during oocyte maturation has demonstrated that immature oocytes contain a monomeric 35 kDa protein Cdc2 but not cyclin B. In oocyte cytoplasm *de novo* synthesis of cyclin B occurs through translational activation by getting the signal from the membrane bound G-protein (α_i) coupled receptor, mPRα, as result of low cyclic AMP (cAMP) and protein kinase A (PKA) activity (Nagahama and Yamashita, 2008). After formation a complex between newly synthesized cyclin B with the pre-existing cdc2 leads to the activation of cdc2 by its phosphorylation at the position Thr161 residue by cdk-activating kinase (CAK). This activation of cdc2 leads to its mobility shift from 35 kDa to 34 kDa protein (Nagahama and Yamashita, 2008). The role of melatonin in steroid induced oocyte maturation was first reported in a study on carp Catla (Chattoraj et al., 2005). In *in vitro* study, denuded oocytes obtained from gravid female carp during spawning phase, incubated with melatonin 4 h prior to the administration of MIH in the medium interestingly increased the rate of GVBD by the formation of MPF (Chattoraj et al., 2005) (Figure 4). Researchers also observed that cyclin B levels are increased gradually in melatonin pre-treated (before 4 h of MIH incubation) oocytes in time dependent manner and reached to its maximum peak after 12 h of incubation. Moreover, the activity of MPF also determined by *H1 kinase* activity assay as a positive indicator of MPF formation and the results significantly specify that more H1 phosphorylation takes place in melatonin pre-treated oocytes in compare to incubation of oocytes with MIH alone. This novel study is a breakthrough to reveal the extra-hypothalamic action of melatonin in the maturation of oocytes in fish for the first time (Chattoraj et al., 2005). The hypothesis earns further support from another study on zebrafish (*Danio rerio*) in which melatonin is known to exert a direct action on follicles (as shown by the increase of the oocytes undergoing to GVBD) and modulates *mpr α* and *mpr β* genes (whose proteins are involved in oocyte maturation) expression (Carnevali et al., 2011). Daily exposure of melatonin with two different doses (100 nM and 1 μM) in tank water to adult females results an increase in gonadosomatic index (GSI) which is associated with the increase in egg production, vitellogenin content (VTG) at both gene and protein levels, and also in the expression of estradiol receptor α (ERα) in the liver. It is also reported that exogenous melatonin leads to the activation of kisspeptin and the production of GnRH (Carnevali et al., 2011). Studies on killifish (*Fundulus heteroclitus*) also provided evidence that melatonin administration (100 nM and 1 μM) for 8 days plays a role in steroidogenesis by regulating the steroidogenic enzymes and their activities in ovary (Lombardo et al., 2012) (Table 2). Melatonin exposure for eight days significantly increased the rate of fecundity and embryo survival in killifish in compare to the control. During follicular maturation, melatonin receptor gene expression is also up-regulated in this marine fish species. Melatonin also administered to the ovarian follicles of killifish in an *in vitro* experiment and results showed that melatonin treated follicles exhibit high percentage of GVBD (Lombardo et al., 2012). Data obtained from the above mentioned various carefully controlled studies it can be concluded that endogenously synthesized or exogenous melatonin play a critical role in the regulation of seasonal reproduction by promoting oocyte maturation in fish.

6.2. In Male

6.2.1. Melatonin and Its Role in Testicular Growth and Development

Melatonin secreted from fish pineal organ plays an important role in testicular growth and development but the studies are fragmentary, and the results are often contradictory as because the effects may vary between species, photoperiodic regimens, and reproductive status of the concerned fish (Bhattacharyya and Maitra, 2006; Bhattacharya et al., 2007). It was reported in Indian major carp (*Catla catla*) that melatonin regulates annual testicular cycle by its pro-gonadal and anti-gonadal effects (Bhattacharya et al., 2007). But it is not yet clearly understood that melatonin is synthesized locally within testes or it comes from the pineal organ via circulation in fish. Administration of exogenous melatonin at different doses (25-, 50-, and 100 μg/100 g body weight/day) for consecutive 15 and 30 days in an adult subtropical male carp *Catla catla* provided evidence for the first time that melatonin regulates annual testicular growth in a dose and duration dependent manner (Bhattacharya et al., 2007). Exogenous melatonin administration, at a dose of 100 μg/100 g body weight for consecutive 30 days resulted into precocious testicular maturation during the preparatory phase, increased the values of both the GSI and the percentage of lobules with mature stage-II spermatids. During this phase, the testicular activity of both $\Delta^5$3β-hydroxysteroid dehydrogenase ($\Delta^5$3β-HSD) and 17β-hydroxysteroid dehydrogenase (17β-HSD) as well as the serum testosterone levels are significantly elevated following the treatment with each dose of melatonin for 30 days. While testicular lobules in control fish contained spermatocytes as the most advanced germ cells. During pre-spawning and spawning seasons, prolonged treatment of exogenous melatonin at a dose of 100 μg/100 g body weight caused an anti-gonadal effect by showing significant variations of both spermatogenic and steroidogenic activity of the testis. On the contrary of preparatory phase, exogenous melatonin resulted in decrease in the value of

GSI, serum testosterone level, the percentage of lobules containing stage-II spermatids, testicular activity of $\Delta^5 3\beta$-HSD and 17β-HSD, in contrast to the testicular lobules showing a large number of stage-II spermatids in control fish (Bhattacharya et al., 2007). Another study on the effects of exogenously administered melatonin reported that dietary supplementation of melatonin at doses 50 or 250 mg/kg balanced fish food for 8 weeks to a male brood stock (first pubertal stage) of walking catfish, *Clarias macrocephalus* exhibited an increase of mean gonadosomatic index, viability rate and progressive motility of the sperm (Aripin et al., 2015). The results have also shown that due to application of high content melatonin with fish food (250 mg/kg fish food), mature spermatozoa are more prevalent than immature spermatogonium as compared to the control group used for the study. However, the results significantly indicate that melatonin can increase the rate of testicular maturation and reduce sperm abnormality which ultimately improves the quality of male seeds (Aripin et al., 2015) (Table 2).

7. Implications of Photoperiods in Seasonal Reproduction of Fish

Duration and intensity of light-dark signal plays a vital role in the regulation of reproduction as it is considered as one of the major external cues through which aquatic vertebrates like fish can synchronize their reproductive events with the environment. Experimental studies in different fish revealed that proper synchronization of physiological functions such as reproduction with the environment is moderated through a system composed of sensors and circadian oscillators like the pineal organ, the lateral eyes and the suprachiasmatic nuclei of the hypothalamus. Pineal organ is considered as the central key component of the circadian oscillating system for its responsiveness due to the changes in environmental light-dark cycle. The pineal hormone melatonin acts as the internal chemical messenger to provide information about the changes in the environmental photoperiodic conditions (Falcón, 1999; Seth and Maitra, 2010). Studies performed in a number of fish species in different photoperiodic conditions found that long photoperiod stimulates ovarian growth in *Heteropneustes fossilis* but short photoperiodic condition significantly inhibits ovarian development (Sundararaj, 1981). The result is also supported by the increment of gonadal growth in *Mystus tengara* (Guraya et al., 1976) and *Channa punctatus* (Srivastava and Singh, 1992) under the exposure of long photoperiods (14L:10D or 18L:06D) as well as continuous illumination (LL) (Srivastava and Singh, 1993). The inhibitory effects on gonadal functions of short photoperiods on is also observed in several fish species including *Cirrhina reba* (Verghese, 1976), *Heteropneustes fossilis* (Sundararaj, 1981), and *Channa punctatus* (Srivastava and Singh, 1992). Long photoperiods in combination with high water temperature, another important environmental cue, effectively stimulate gonadal maturation in *Channa punctatus* (Srivastava and Singh, 1992). Experimental studies on Indian major carp *Catla catla* also revealed that long photoperiod (LP: 16L:08D) enhances steroidogenic properties of the ovary ($\Delta^5$3β-HSD and 17β-HSD) and serum VTG protein content during preparatory phase but short photoperiod (SP: 08L:16D) has no influence on the ovarian activity in this season (Dey et al., 2005). The result interestingly explained that LP condition caused precocious maturation during pre-spawning phase by increasing both steriodogenic ($\Delta^5$3β-HSD and 17β-HSD) and gametogenic (GSI value, more percentage of mature oocytes) functions of the gonads in compare to control as well as SP conditions. Moreover, during spawning phase, LP has no impacts on reproductive functions but SP exhibits an inhibition to ovarian growth and maturation. The post-spawning phase in an annual ovarian cycle of fish is irresponsive to both LP and SP conditions (Dey et al., 2005). In a separate experiment on the same fish species (Catla), results demonstrated that continuous light (LL) and continuous darkness (DD) have reproductive phase dependent responsiveness in an annual testicular cycle (Bhattacharya et al., 2007). During preparatory phase, continuous darkness (DD) caused precocious maturation of testicular functions. On the contrary, during pre-spawning phase LL led to the stimulation of the testicular maturation and milt-oozing conditions (Bhattacharya et al., 2007). Comparable data are also obtained from various experimental studies, on red sea bream (*Pagrus major*), Nile tilapia (*Oreochromis niloticus*), haddock (*Melanogrammus aeglefinus*), and black sea turbot (*Psetta maeotica*), demonstrated that fish exposed to continuous light (24L:00D) significantly lowers gonadal growth as compare to fish exposed to different long photoperiodic manipulations (20L:04D, 18L:06D, 16L:08D) (Trippel and Neil, 2003; Türker et al., 2005; Biswas et al., 2006). Studies on *Oreochromis niloticus* also reported that constant light (24L:00D) significantly lowered the mean gonosomatic index (GSI) and oocyte size in fish as compared to those maintained under long (20L:04D, 18L:06D) photoperiodic conditions and control (Rad et al., 2006) (Table 2). Thus, melatonin acts as an endocrine messenger of the photoperiodic signal which seems to be involved in the regulation of seasonal reproduction and the functional responsiveness of this hormone depends upon the sexual status of the fish (Maitra et al., 2015). The discussed observations on the role of different photoperiodic schedules in fish species appears that knowledge on the mechanisms involved in the processing of light signal in controlling the reproductive events may provide an enormous progress in the field of aquaculture.

8. Molecular Mechanism of Action of Melatonin on Fish Reproduction

8.1. Action of Melatonin on Hypothalamo-Hypophyseal-Gonadal (HPG) Axis

The role of melatonin in the regulation of reproduction is one of the major areas that received wide attention for different non-mammalian animal species. In different animal species, the hypothalamo-hypophyseal system plays a major role in the photoperiodic regulation of reproduction, where melatonin provides temporal information related to day length and season (Falcón et al., 2007). *In vitro* pharmacological investigations clearly demonstrated that melatonin modulates the neuroendocrine functions for regulating seasonal reproduction by targeting the melatonin receptors, as detected by PCR in different components of the HPG axis (Falcón et al., 2003). Melatonin acts through two types of membrane bound G-protein coupled receptor isoforms, MT1 and MT2, and this hormone also acts through a nuclear receptor from the retinoic acid orphan receptors family, ROR/RZR (Richter et al., 2004). The MT1 and MT2 receptors are widely distributed in various tissues and organs such as retina, pars tuberalis (PT) of pituitary gland, suprachiasmatic nucleus (SCN) of the hypothalamus, cerebellum, hippocampus, kidney, intestine, lungs, ovary, testis, mammary glands, adrenal gland, lymphocytes, gall bladder, skin, adipocytes, blood vessels (Klosen et al., 2019). These two receptor subtypes vary in terms of their affinity for melatonin, cellular distribution and the involvement of second messengers in signalling pathway. Another third type of melatonin receptor, MT3, has also been found which acts like quionone reductase (QR2) which might be involved in mediating the antioxidant effects of melatonin (Pfeffer et al., 2017). MT3 is not considered as classic melatonin membrane receptor and several reports indicate that it is located within the cytosol rather than cell membranes (Tan et al., 2014). These three receptor subtypes were previously known as Mel1a, Mel1b and Mel1c. In mammals, two melatonin receptors MT1 and MT2 are found whereas in non-mammalian species three receptor subtypes are found (Pevet, 2000). The circadian clock SCN of the hypothalamus along with the PT of anterior pituitary contain the highest density of melatonin receptors (Gorman, 2020) and this spatial distribution of melatonin receptor is believed to be involved in the regulation of circadian rhythms and reproductive functions (von Gall et al., 2002). The localization of melatonin receptors in fish whole brain is demonstrated by Shi et al., (2004). Melatonin receptors and their mRNA expressions in the brain and retina also show temporal variations along with the environmental light-dark conditions in processing the photoperiodic information (Shin et al., 2011). Studies performed within the last few years explained that melatonin exerts its direct effects in fish ovary through the activation of melatonin receptors Mel1b (MT1) and Mel1b (MT2) proteins (Chattoraj et al., 2009; Moniruzzaman and Maitra, 2012). In Indian major carp *Catla catla*, 37 kDa Mel1aR (MT1) receptor protein is detected through Western blot analysis in the membrane and cytosolic segments of the ovary in carp provided evidence that melatonin may have a direct role in oocyte maturation in a receptor dependent manner (Chattoraj et al., 2009). Scientists also showed the mRNA expression of melatonin receptors (MT1, MT2, and MT3) in mudskipper (*Boleophthalmus pectinirostris*) (Hong et al., 2014), orange-spotted grouper (*Epinephelus coioides*) (Chai et al., 2013), Nile tilapia (*Oreochromis niloticus*) (Jin et al., 2013), and in medaka (*Oryzias latipes*) (Ogiwara and Takahashi, 2016). Studies on the carp *Catla catla*, the temporal pattern of ovarian MT1 and MT2 receptor proteins significantly varied in relation to the reproductive status of the fish in an annual cycle (Moniruzzaman and Maitra, 2012). The protein density in the ovary, a component of HPG axis, is maximum during the spawning and lowest during post-spawning phase.

Obviously, melatonin *per se* is neither anti-gonadotrophic nor pro-gonadotrophic, rather the changing duration of the nocturnal melatonin message is a passive signal that provides information on the time of year (calendar information) to HPG axis (Hazlerigg, 2001). The reproductive axis uses the season related melatonin rhythm to adjust time of reproduction accordingly. However, considering hypothalamus as the highest centre of regulation of pituitary functions, more convincing data on the mechanism of melatonin action on this part of brain deserves special attention. Melatonin has been implicated in regulation of reproduction on HPG axis by the synthesis or release of hypothalamic GnRH, kisspeptins or GnIH, but the mechanism of their interaction still remains to be specified properly (Takahashi and Ogiwara, 2021). Existence of day–night variations as well as melatonin inhibitory effects on the expression of GnRH-1, GnRH-3 and several GnRH receptors in the brain are reported in European sea bass (Servili et al., 2013). Obviously, these interactions between melatoninergic and GnRH systems could represent a substrate of photoperiod effects on reproductive and other rhythmic physiological events in European sea bass and other fish species, in which melatonin is directly implicated to photoperiodic regulation of seasonal reproduction. Further studies provide evidence that melatonin induces GnIH expression in GnIH neurons (Tsutsui et al., 2012). Due to the expression of melatonin receptor (Mel1c) in the GnIH neurons, melatonin may act directly on the neurons to induce GnIH expression and release (Tsutsui, 2009). An *in vitro* study on the effects of melatonin demonstrated a dose-dependent increase in GnIH release from the hypothalamus (Chowdhury et al., 2010). The expression of GnIH in brain is increased under short day, when the nocturnal duration of melatonin release is also increased, indicating that melatonin may directly induce synthesis of GnIH (Ubuka et al., 2005). GnIH via altered melatonin signal may transduce photoperiodic information to the avian reproductive axis (Tsutsui et al., 2012). However, it remains to be known whether such mechanism also operates in seasonally reproducing teleosts, including carp (Maitra et al., 2013). Likewise, the functional interplay between the GnRH, GnIH and melatonin at hypothalamic level in controlling reproduction in seasonally breeding fish is far from being completely resolved.

8.2. Direct Action of Melatonin as an Antioxidant in Fish Reproduction

Due to the universal occurrence of melatonin in every possible living species so far investigated, it is considered to be a phylogenetically ancient molecule and it was probably evolved as a scavenger of oxygen free radicals and during the course of evolution it acquired different functions within the living systems (Reiter et al., 2013). A novel finding, that melatonin, due to its amphiphilic nature, can easily cross the cellular barrier and may act as a highly ubiquitous direct free radical scavenger and indirect antioxidant (Tan et al., 1993), has opened up a new possibility that this indoleamine, in addition to its receptor-mediated hormonal actions, may perform a receptor-independent action in the regulation of a number of physiological functions, including reproduction (da Silva et al., 2015). Since melatonin is capable of donating electrons to reduce the reactivity of molecules with an unimpaired electron in their valance orbital, i.e., free radicals, it may apparently directly interact with potentially damaging agents without the necessity of first binding to a receptor. It also stimulates a number of enzymes which are closely associated with the production of different endogenous antioxidants (Rodriguez et al., 2004; Maitra and Hasan, 2016). Notably, free-radical scavenging activity of melatonin does not require any receptor at the target cells. Ianăş et al., (1991) provided first indication that melatonin may be a direct free radical scavenger. Two years later, Tan et al., (1993) presented strong evidence that melatonin is highly effective in detoxifying the highly reactive hydroxyl radical (•OH) and a variety of free radicals and reactivates oxygen intermediates (Maitra and Hasan, 2016). The studies demonstrating further that even scavenging actions of the metabolites of melatonin on free-radicals (Galano et al., 2013) added a new dimension to the current understanding on the physiological role of melatonin as a potent antioxidant in the regulation of ovarian functions in at least mammals (Tamura et al., 2020). Studies on several mammalian species including human revealed that melatonin concentrations in the ovary is found to be higher in pre-ovulatory follicular fluid than in the serum, may have an important role during ovulation (Tamura et al., 2020). Since high levels of oxygen free radicals are commonly generated during oocyte maturation and during ovulation, melatonin through its free radical scavenging activity plays a very vital role in the protection of oocytes by reducing the harmful effects of ROS (Tamura et al., 2009). Thus, melatonin through its receptor independent actions in the female reproductive system can directly act as an endogenous antioxidant which is responsible for maintaining ovarian homeostasis and development of good quality oocytes. Study on carp and several other vertebrates also have demonstrated the indirect actions of melatonin in a receptor independent way by increasing the activity of different antioxidative enzymes in the ovary (Hasan et al., 2014; Maitra et al., 2015; Maitra and Hasan, 2016). The development and maturation of oocytes require the synthesis of steroids in the ovary which is commonly associated with the generation of reactive oxygen species (ROS) and reactive nitrogen species (RNS) (Valenzuela et al., 2015) which can lead to the damage of developing oocytes and cause oocyte aging resulting non-viable and low quality of eggs (Olcese, 2020). The ameliorating property of melatonin which is based upon its scavenging activity can reduce free radical induce damages and improves the quality of oocytes (Reiter et al., 2013).

Follicle development in the mammalian ovary is a complex physiological process during which active synthesis of steroids is accompanied by the production of large content of free-radicals, especially reactive nitrogen species (RNS) and reactive oxygen species (ROS) (Tamura et al., 2020). Excessive amount of free radicals, especially ROS, cause oxidative stress resulting in damage of oocytes (Maitra and Hasan, 2016), accelerate oocyte aging and deteriorate oocyte quality (Goud et al., 2008). The study on a carp (*Cyprinus carpio*) also provided indication that steroidogenesis in the ovarian follicular cells is related to generation of a large quantity of free radicals (Behl, 2006). One of the alleged roles of melatonin detected in the mammalian pre-ovulatory follicles is the protection of oocyte against free radicals by playing a potent role in the physiological cascade of oocyte maturation (Acharyya et al., 2021). *In vitro* study on isolated oocytes revealed that elevation of melatonin level in the ovary depends upon follicular growth during the complex process of ovulation (Nakamura et al., 2003; Tamura et al., 2020). Available data also suggest that melatonin may act on various cells of ovarian follicles as an indirect antioxidant to activate major antioxidative enzymes like superoxide dismutase (SOD), catalase (CAT), glutathione peroxidase (GPx), glutathione S-transferase (GST), glutathione reductase (GRd) which metabolize free radicals to reduce oxidative stress *in vivo* (Hasan et al., 2014; 2020). SODs represent a class of closely related enzymes that catalyze the breakdown of the superoxide anion into oxygen and hydrogen peroxide (H_2O_2) (Zelko et al., 2002). GPx and CAT then independently convert H_2O_2 to water (Gaeta et al., 2002). SOD together with GPx and CAT form the main enzyme defense mechanism against harmful effects of free-radicals. GST (glutathione S-transferase), like GPx, is an important enzyme in the glutathione redox cycle catalyzing the conjugation of reduced glutathione via a sulfhydryl group to xenobiotic substrates for detoxification purpose. GRd catalyzes the reduction of glutathione disulfide (GSSG) to the sulfhydryl form glutathione (GSH). Reduced glutathione (GSH), a non-enzymatic tripeptide antioxidative agent, participates in removal of toxic peroxides and aldehydes (Hasan et al., 2020). The hypothesis is now well tested in fish species and the knowledge will be helpful for the production of healthy egg which is one of the major critical factors in induced breeding. Study on carp *Catla catla* under natural photo-thermal conditions revealed for the first time that intra-ovarian melatonin exhibits seasonal periodicity with a peak at spawning phase and nadir at post-spawning phase (Hasan et al., 2014). Researchers also demonstrated that melatonin concentration is positively correlated strongly with the activity of enzymatic (SOD, CAT, GPx, GST) and levels of non-enzymatic (GSH)

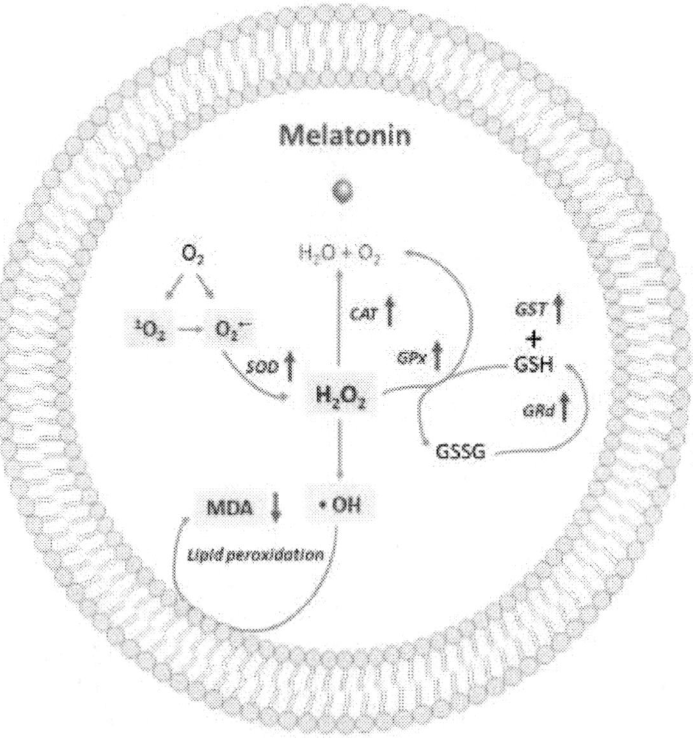

Abbreviations: CAT: Catalase; GPx: Glutathione peroxidase; GRd: Glutathione reductase; GSH: Reduced glutathione; GSSG: Oxidized glutathione; GST: Glutathione S-transferase; MDA: Malondialdehyde; SOD: Superoxide dismutase.

Figure 5. Melatonin as an antioxidant in the liver and ovary of carp *Catla catla*. Schematic presentation demonstrates the possible effects of melatonin on the activity of antioxidative enzymes and different free-radicals [$^1O_2$, $O_2^{\cdot-}$, $OH^{\cdot}$] producing oxygen based stress within the carp hepatocyte and oocyte. Recent experimental results suggest that melatonin reduces (downward arrow) intracellular MDA level by stimulating (upward arrow) the activities of enzymatic (SOD, CAT, GPx, GRd and GST) and non-enzymatic (GSH) antioxidants in both liver and ovary to establish intracellular antioxidative milieu.

antioxidants underlines the functional interplay as well as importance of intra-ovarian melatonin in reducing oxidative stress. However, melatonin showed a significant negative correlation with malondialdehyde (MDA) level, a marker of intracellular oxidative stress, in an annual ovarian cycle indicating that melatonin reduces the oxidative stress to augment ovarian functions during the spawning. In other experiments on the same fish species, exogenous melatonin administration effectively reduced MDA level but various enzymatic and non-enzymatic antioxidants are increased significantly in the ovary in each reproductive phase in an annual gonadal cycle (Moniruzzaman et al., 2016; Mondal et al., 2017). It has also been reported in a very recent study on the same carp *Catla catla* that hepatic melatonin also reduces 'reactive oxygen species' (ROS) generation (as MDA level is decreased) in an annual reproductive cycle through the enhancement of the activities/levels of antioxidant enzymes such as SOD, CAT, GPx, GST, GRd and non-enzymatic antioxidant GSH as observed in the ovary during oocyte growth and maturation in fish (Hasan et al., 2020) (Figure 5). The results also demonstrate the individual role of different antioxidative enzymes in the melatonin-induced ovarian growth and maturation to improve the quality of oocytes (Lombardo et al., 2014; Maitra and Hasan, 2016) in carp as well as in other fish species remains as an exciting topic of future research as the mechanism of free radical scavenging activity of melatonin on the carp oocytes remains speculative.

9. Application of Melatonin as a 'Hormonal Tool' in Aquaculture

Melatonin may also be used as a 'hormonal tool' for induced breeding practices in aquaculture for its ameliorative potential in reproductive physiology in fish. The hypothesis earned a strong support from the results of a recent *in vivo* study in adult female carp *Catla catla* in which the process of final oocyte maturation (FOM) and oxidative status of the pre-ovulatory follicles are studied (Moniruzzaman et al., 2016). Fish were injected with melatonin (25 μg/100 g body weight) and/or ovaprim (1 mL containing 20 mg salmon GnRH analogue or sGnRHa and 10 mg domperidone dissolved in propylene glycol), or luzindole (a pharmacological blocker of melatonin receptors) (0.1 μg/100 g body weight) before them. The effects of melatonin on latency period, time in between the last hormone injection and occurrence of ovulation, is calculated. The latency period is longest (11.80 ± 0.74 h) in luzindole pre-treated, ovaprim injected fish and shortest (7.30 ± 0.22 h) in melatonin treated fish, 2 h before the primary dose (0.15 mL/kg body weight) of ovaprim, as compared to the latency period (10.47 ± 0.59 h) of only ovaprim treated carp. Melatonin has also exerted a positive effect on the GVBD, another important index of FOM, that, melatonin injection 2 h before of the 1st dose of ovaprim leads to maximum rate (96.3 ± 3.6%) of GVBD in compare to (26.3 ± 2.4%) only melatonin treated carp ovary. Another important index of FOM is the rate of formation of active MPF. The results significantly highlight that melatonin pre-treatment (2 h before of ovaprim 1st dose) leads to the formation of active maturation promoting factor (MPF), a complex of cyclin B and cyclin dependent kinase 1 (CDK1), and increases the rate of ovulation as the latency period becomes shorten under natural photo-thermal conditions (Figure 6). Exogenous melatonin, irrespective of injection schedule, leads to a considerable reduction in intra-follicular oxidative stress and an increase in the levels of different ovarian antioxidants, melatonin, and its membrane bound MT1 receptor protein. Interestingly, due to treatment of luzindole, an inhibitor of melatonin receptors, no significant effects of melatonin and/or ovaprim is observed and thus the study provided the evidence that melatonin can ameliorate the effects of ovaprim actions on oocyte maturation in fish (Moniruzzaman et al., 2016). Malondialdehyde (MDA), an intracellular stress marker, is significantly reduced in the melatonin pre-treated carp, while MDA levels are significantly increased in luzindole pre-treated carp ovary. The activity of antioxidative enzymes, such as superoxide dismutase (SOD), catalase (CAT), glutathione peroxidase (GPx), gluathione S-transferase (GST) and glutathione reductase (GRd) and non-enzymatic antioxidantive agent such as reduced glutathione (GSH) are significantly increased in the melatonin treated carp.

A correlative study in this experiment also suggests that a significant negative correlation between ovarian melatonin and the level of intra-ovarian oxidative stress and a positive correlation with the rate of GVBD and the levels/activity of different antioxidative enzymes are the critical synchronizers in FOM. The band intensity of MT1 receptor protein in the samples is always higher in melatonin treated fish (alone or along with or 2 h before the treatment of ovaprim at the 1st dose) than in the respective control or booster dose of ovaprim (0.35 mL/kg body weight) (with or without melatonin) treatment or luzindole-pre-treated melatonin injected carp. Thus, it can be suggested that treatment with exogenous melatonin can promote oocyte maturation through the reduction of intracellular oxidative stress and by stimulating the antioxidative defence mechanism in ovary (Moniruzzaman et al., 2016) (Figure 6). Another approach of the use of melatonin in aquaculture is the dietary supplementation of melatonin has an impact in aquaculture as it plays a positive role in final oocyte maturation and in the generation of high-quality oocytes (Acharyya et al., 2021). Data obtained from different studies provide evidence that dietary supplementation of melatonin or by providing tryptophan rich diet may promote the synthesis of melatonin in the gut or can modulate the diurnal profiles of melatonin synthesis (Peuhkuri et al., 2012). Thus, through the regulation of feeding time and supplementation of tryptophan-rich food may stimulate the synthesis of melatonin in the gut which can ultimately be released into the blood to contribute the total pool of circulating levels of melatonin (Pal et al., 2016a). Further studies on the implications of melatonin as a tool in aquaculture are thus required for the better understanding of the role of melatonin in the modulation of physiological properties including reproduction in fish.

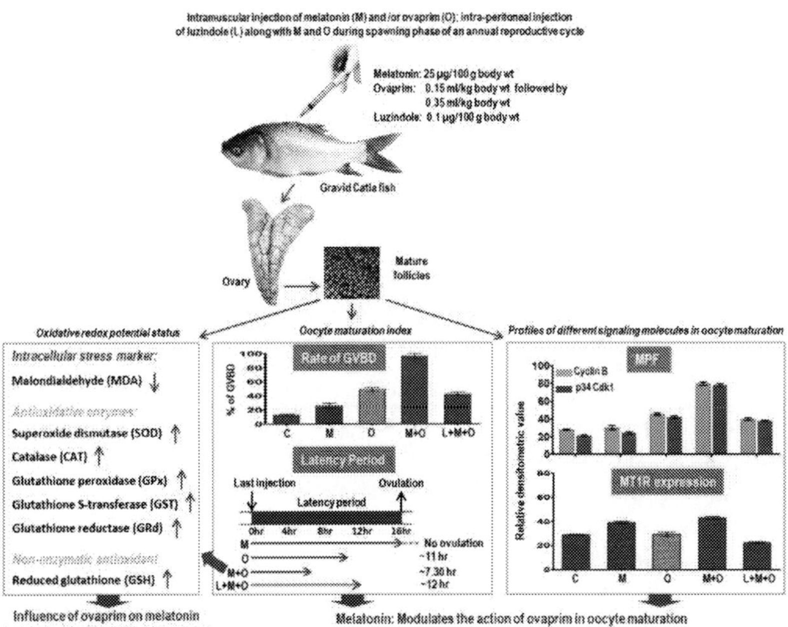

Experimental combinations: C: Control; M: Only melatonin; O: Only ovaprim; M+O: Melatonin (–2 h) and ovaprim; L+M+O: Luzindole, melatonin (–2 h) and ovaprim.

Figure 6. Implication of melatonin on ovaprim-induced oocyte maturation. Diagrammatic and graphical presentations based on the data obtained from a recent study on carp *Catla catla*. Maturational and antioxidative effects of exogenous melatonin (25 µg/100 g body) in different combinations with ovaprim as well as in presence or absence of luzindole has been evaluated in the mature ovarian follicles. Melatonin, 2 h prior to the administration of ovaprim, has significant potential role in the reduction of the latency period, induction of GVBD by promoting the synthesis and activation of maturation promoting factor (MPF) and MT1 receptor protein, maintenance of intra-follicular redox balance by reducing intracellular stress and stimulating the activities/levels of enzymatic and non-enzymatic antioxidants.

Conclusion

Melatonin has many implications in various physiological functions among which regulation of reproduction is an important part which draws the attention of researchers around the world as it may aid in the development of its practical application in aquaculture. The past few decades have witnessed an enormous progress in research in the field of fish reproductive biology and endocrinology but demonstration of the proper mechanism of its action is not yet clearly elucidated. Melatonin being a multipotent endogenous chronobiotic molecule is responsible in the regulation of fish reproduction and thus its application in aquaculture through various carefully controlled routes may not be ruled out to enhance the production rate of fish to fulfil its increasing demand. In fish, as a seasonal breeder, environment especially photoperiod is a very vital limiting factor in reproduction. Melatonin as a critical messenger of the light-dark signal, it can synchronize fish spawning with a specific favourable time of the year. So, the manipulations of photoperiodic conditions may also pertain as a factor of fish reproduction by changing the circulating as well as intra-ovarian levels of melatonin. Thus, the basic knowledge on the pleiotropic action of melatonin evidenced from both *in vivo* as well as *in vitro* studies may be used as an effective agent in aquaculture practices for obtaining improved quality of male and female seeds and an increased rate of reproduction but further studies are still required on the beneficial role of melatonin in reproductive performance in field-based culture system.

Acknowledgments

Financial assistance from University Grants Commission (UGC) [F.30-448/2018(BSR)] to KNH is thankfully acknowledged. Swami Vivekananda Merit cum Means Fellowship (WBP191602488458), Govt. of West Bengal, India to JD and the award of Senior Research Fellow (SRF) [F. No. 09/1156(0006)/2018-EMR-I] to AA from the Council of Scientific and Industrial Research (CSIR), Govt. of India, is appreciatively acknowledged.

References

Acharyya, Akash, Joydeep Das, and Kazi Nurul Hasan, 2021. "Melatonin as a multipotent component of fish feed: basic information for its potential application in aquaculture." *Frontiers in Marine Science*, 8:1298. doi:10.3389/fmars.2021.734066.

Acuña-Castroviejo, Darío, Germaine Escames, Carmen Venegas, María E. Díaz-Casado, Elena Lima-Cabello, Luis C. López, Sergio Rosales-Corral, Dun-Xian Tan, and Russel J. Reiter. 2014. "Extrapineal melatonin: sources, regulation, and potential functions." *Cellular and molecular life sciences* 71:2997-3025. doi:10.1007/s00018-014-1579-2.

Akashi, Makoto, and Toru Takumi. 2005. "The orphan nuclear receptor RORα regulates circadian transcription of the mammalian core-clock Bmal1." *Nature structural & molecular biology* 12:441-448. doi:10.1038/nsmb925.

Akashi, Makoto, Yoshiki Tsuchiya, Takao Yoshino, and Eisuke Nishida. 2002. "Control of intracellular dynamics of mammalian period proteins by casein kinase I ε (CKIε) and CKIδ in cultured cells." *Molecular and cellular biology* 22:1693-1703. doi:10.1128/MCB.22.6.1693-1703.2002.

Albrecht, Martin. 2009. "Insights into the nature of human testicular peritubular cells." *Annals of Anatomy-Anatomischer Anzeiger* 191:532-540. doi:10.1016/j.aanat.2009.08.002.

Amano, Masafumi, Masayuki Iigo, Kazumasa Ikuta, Shoji Kitamura, Hideaki Yamada, and Kunio Yamamori. 2000. "Roles of melatonin in gonadal maturation of underyearling precocious male masu salmon." *General and comparative endocrinology* 120:190-197. doi:10.1006/gcen.2000.7547.

Amaral, Fernanda Gaspar do, and José Cipolla-Neto. 2018. "A brief review about melatonin, a pineal hormone." *Archives of endocrinology and metabolism* 62: 472-479. doi:10.20945/2359-3997000000066.

Aripin, Siti-Ariza, Orapint Jintasataporn, and Ruangvit Yoonpundh. "Effects of exogenous melatonin in *Clarias macrocephalus* male broodstock first puberty stage. 2015. " *Journal of Aquaculture Research & Development* 6:1. doi:10.4172/2155-9546.1000307.

Axelrod, Julius, and Herbert Weissbach. 1961. "Purification and properties of hydroxyindole-O-methyl transferase." *Journal of Biological Chemistry* 236: 211-213. doi:10.1016/S0021-9258(18)64458-8.

Bayarri, María José, Silvia Zanuy, Ozlem Yilmaz, and Manuel Carrillo. 2009. "Effects of continuous light on the reproductive system of European sea bass gauged by alterations of circadian variations during their first reproductive cycle." *Chronobiology International* 26:184-199. doi:10.1080/07420520902758311.

Bégay, Valérie, Jack Falcón, Gregory M. Cahill, David C. Klein, and Steven L. Coon.1998. "Transcripts encoding two melatonin synthesis enzymes in the teleost pineal organ: circadian regulation in pike and zebrafish, but not in trout." *Endocrinology* 139:905-912. doi:10.1210/endo.139.3.5790.

Behl, Rahul. 2006. "Stimulation of Catalase Activity in Carp (*Cyprinus carpio*) Ovarian Follicles by Gonadotropin *in vitro*." *Fish Physiology and Biochemistry* 32:203-207. doi:10.1007/s10695-006-0012-7.

Bellingham, James, Shyam S. Chaurasia, Zara Melyan, Cuimei Liu, Morven A. Cameron, Emma E. Tarttelin, P. Michael Iuvone, Mark W. Hankins, Gianluca Tosini, and Robert J. Lucas. 2006. "Evolution of melanopsin photoreceptors: discovery and characterization of a new melanopsin in nonmammalian vertebrates." *PLoS biology* 4:e254. doi:10.1371/journal.pbio.0040254.

Besseau, Laurence, Ahmed Benyassi, Morten Møller, Steven L. Coon, Joan L. Weller, Gilles Boeuf, David C. Klein, and Jack Falcón. 2006. "Melatonin pathway: breaking the 'high-at-night'rule in trout retina." *Experimental eye research* 82:620-627. doi:10.1016/j.exer.2005.08.025.

Bhattacharya, S. 1999. "Recent advances in the hormonal regulation of gonadal maturation and spawning in fish." *Current science* 342-349.

Bhattacharya, S., R. Dey, A. Basu, S. K. Maitra, and T. K. Banerji. 2003. "The structure of the pineal complex in a common Indian teleost, *Catla catla*: evidence for pineal-induced inhibition of testicular function within an annual reproductive cycle." *Endocrine research* 29:141-156. doi:10.1081/ERC-120022295.

Bhattacharya, Sharmistha, Asamanja Chattoraj, and Saumen Kumar Maitra. 2007. "Melatonin in the regulation of annual testicular events in carp *Catla catla*: evidence from the studies on the effects of exogenous melatonin, continuous light, and continuous darkness." *Chronobiology international* 24:629-650. doi:10.1080/07420520701534665.

Bhattacharyya, Sharmistha, Ruma Dey, and Saumen Kumar Maitra. 2005. "Photoperiodic regulation of annual testicular events in the Indian major carp *Catla catla*." *Acta Zoologica* 86:71-79. doi:10.1111/j.1463-6395.2005.00188.x.

Bhattacharyya, Sharmistha, and Saumen Kumar Maitra. 2006. "Environmental correlate of the testicular events in a major carp *Catla catla* in an annual reproductive cycle." *Biological Rhythm Research* 37:87-110. doi:10.1080/09291010500124605.

Biswas, Amal K., Manabu Seoka, Yoshimasa Tanaka, Kenji Takii, and Hidemi Kumai. 2006. "Effect of photoperiod manipulation on the growth performance and stress response of juvenile red sea bream (*Pagrus major*)." *Aquaculture* 258:350-356. doi:10.1016/j.aquaculture.2006.03.048.

Blasiak, Janusz, Russel J. Reiter, and Kai Kaarniranta. 2016. "Melatonin in retinal physiology and pathology: the case of age-related macular degeneration." *Oxidative medicine and cellular longevity*. 2016. doi:10.1155/2016/6819736.

Bornestaf, Cecilia, Ian Mayer, and Bertil Borg. 2001. "Melatonin and maturation pace in female three-spined stickleback, Gasterosteus aculeatus." *General and comparative endocrinology* 122:341-348. doi:10.1006/gcen.2001.7640.

Bromage, N. R., C. Randall, B. Davies, and M. Porter. 1994. "Photoperiod, melatonin and fish reproduction." In *High Performance Fish: Proceedings Of An International Fish Physiology Symposium At The University Of British Columbia*, edited by Mackinlay DD, 137-142. Canada: Fish Physiology Association.

Bromage, N. R., C. F. Randall, M. J. R. Porter, and B. Davies. 1995. "How do photoperiod, the pineal gland, melatonin, and circannual rhythms interact to co-ordinate seasonal reproduction in salmonid fish." In *Proceedings of the Fifth International Symposium on the Reproductive Physiology of Fish, The University of Texas at Austin*, edited by Goetz F, and Thomas O., 164-166.

Bromage, Niall, Mark Porter, and Clive Randall. 2001. "The environmental regulation of maturation in farmed finfish with special reference to the role of photoperiod and melatonin." *Aquaculture* 197:63-98.

Brzezinski, Amnon, Machelle M. Seibel, Harry J. Lynch, Mei-Hua Deng, and Richard J. Wurtman. 1987. "Melatonin in human preovulatory follicular fluid." *The Journal of Clinical Endocrinology & Metabolism* 64:865-867. doi:10.1210/jcem-64-4-865.

Brzezinski, Amnon, Seema Rai, Adyasha Purohit, and Seithikurippu R. Pandi-Perumal. 2021. "Melatonin, Clock Genes, and Mammalian Reproduction: What Is the Link?." *International Journal of Molecular Sciences* 22:13240. doi:10.3390/ijms222413240.

Bubenik, George A. 1980. "Localization of melatonin in the digestive tract of the rat." *Hormone Research in Paediatrics* 12:313-323. doi:10.1159/000179137.

Bubenik, George A. 2002. "Gastrointestinal melatonin: localization, function, and clinical relevance." *Digestive diseases and sciences* 47:2336-2348. doi:10.1023/A:1020107915919.

Bubenik, George A., and Gregory M. Brown. 1997. "Pinealectomy reduces melatonin levels in the serum but not in the gastrointestinal tract of rats." *Neurosignals* 6:40-44. doi:10.1159/000109107.

Bubenik, George A., and Shiu Fun Pang. 1997. "Melatonin levels in the gastrointestinal tissues of fish, amphibians, and a reptile." *General and Comparative Endocrinology* 106:415-419. doi:10.1006/gcen.1997.6889.

Bubenik, G. A., R. A. Purtill, G. M. Brown, and, L. J. Grota. 1978. "Melatonin in the retina and the Harderian gland. Ontogeny, diurnal variations and melatonin treatment." *Experimental eye research* 27:323-333. doi:10.1016/0014-4835(78)90166-5.

Bubenik, George A., Ball Ronald O., and Pang Shiu-Fun. 1992. "The effect of food deprivation on brain and gastrointestinal tissue levels of tryptophan, serotonin, 5-hydroxyindoleacetic acid, and melatonin." *Journal of pineal research* 12:7-16. doi:10.1111/j.1600-079X.1992.tb00020.x.

Bubenik, G. A., S. F. Pang, R. R. Hacker, and P. S. Smith, 1996. "Melatonin concentrations in serum and tissues of porcine gastrointestinal tract and their relationship to the intake and passage of food." *Journal of pineal research* 21:251-256. doi:10.1111/j.1600-079X.1996.tb00294.x.

Carnevali, O., G. Gioacchini, C. C. Piccinetti, F. Maradonna, F. Lombardo, E. Giorgini, and G. Tosi. 2010. "Melatonin control of oogenesis and metabolic resources in Zebrafish." *Journal of Applied Ichthyology* 26:826-830. doi:10.1111/j.1439-0426.2010.01549.x.

Carnevali, Oliana, Giorgia Gioacchini, Francesca Maradonna, Ike Olivotto, and Beatrice Migliarini. 2011. "Melatonin induces follicle maturation in Danio rerio." *PLoS One* 6:e19978. doi: 10.1371/journal.pone.0019978.

Chai, K. E., Xiaochun Liu, Yong Zhang, and Haoran Lin. 2013. "Day-night and reproductive cycle profiles of melatonin receptor, kiss, and gnrh expression in orange-spotted grouper (Epinephelus coioides)." *Molecular reproduction and development* 80:535-548. doi:10.1002/mrd.22191.

Chattoraj, Asamanja, Sharmistha Bhattacharyya, Dipanjan Basu, Shelley Bhattacharya, Samir Bhattacharya, and Saumen Kumar Maitra. 2005. "Melatonin accelerates maturation inducing hormone (MIH): induced oocyte maturation in carps." *General and Comparative Endocrinology* 140:145-155. doi:10.1016/j.ygcen.2004.10.013.

Chattoraj, Asamanja, Mohua Seth, and Saumen Kumar Maitra. 2009. "Localization and dynamics of Mel1a melatonin receptor in the ovary of carp *Catla catla* in relation to serum melatonin levels." *Comparative Biochemistry and Physiology Part A: Molecular & Integrative Physiology* 152:327-333. doi:10.1016/j.cbpa.2008.11.010.

Chowdhury, Indrajit, and Saumen Kumar Maitra. 2012. "Melatonin time line: From discovery to therapy." *In Melatonin in the Promotion of Health Chapter 1, Second Edition*, edited by R R Watson, 1-60. Taylor & Francis Group.

Chowdhury, Vishwajit S., Kazutoshi Yamamoto, Takayoshi Ubuka, George E. Bentley, Atsuhiko Hattori, and Kazuyoshi Tsutsui. 2010. "Melatonin Stimulates the Release of Gonadotropin-Inhibitory Hormone by the Avian Hypothalamus." *Endocrinology* 151:271-280. doi:10.1210/en.2009-0908.

da Silva, Danilo Grünig Humberto, Octavio Ricci Jr, Eduardo Alves de Almeida, and Claudia Regina Bonini-Domingos. 2015. "Potential utility of melatonin as an antioxidant therapy in the management of sickle cell anemia." *Journal of pineal research* 58:178-188. doi:10.1111/jpi.12204.

Davie, Andrew, Carlos Mazorra de Quero, Niall Bromage, Jim Treasurer, and Herve Migaud. 2007. "Inhibition of sexual maturation in tank reared haddock (*Melanogrammus aeglefinus*) through the use of constant light photoperiods." *Aquaculture* 270:379-389. doi:10.1016/j.aquaculture.2007.04.052.

Devi, Haobijam Sanjita, Chongtham Rajiv, Gopinath Mondal, Zeeshan Ahmad Khan, Sijagurumayum Dharmajyoti Devi, Thangal Yumnamcha, Rupjyoti Bharali, and Asamanja Chattoraj. 2016. "Melatonin bio-synthesizing enzyme genes (Tph1, Aanat1, Aanat2, and Hiomt) and their temporal pattern of expression in brain and gut of a tropical carp in natural environmental conditions." *Cogent Biology* 2:1230337. doi:10.1080/23312025.2016.1230337.

Dey, Ruma, S. Bhattacharya, S. K. Maitra, and T. K. Banerji. 2003. "The morpho-anatomy and histology of the pineal complex in a major indian carp, *Catla catla*: Identification of the pineal photoreceptor cells and their responsiveness to constant light and constant darkness during different phases of the annual reproductive cycle." *Endocrine research* 29(4): 429-443. doi:10.1081/ERC-120026949.

Dey, Ruma, Sharmistha Bhattacharya, and Saumen Kumar Maitra. 2004. "Temporal pattern of ovarian activity in a major carp *Catla catla* and its possible environmental correlate in an annual cycle." *Biological Rhythm Research* 35:329-353. doi:10.1080/09291010400003792.

Dey, Ruma, Sharmistha Bhattacharya, and Saumen Kumar Maitra. 2005. "Importance of photoperiods in the regulation of ovarian activities in Indian major carp *Catla catla* in an annual cycle." *Journal of biological rhythms* 20:145-158. doi:10.1177/0748730404272925.

Devi, Sijagurumayum Dharmajyoti, Gopinath Mondal, Zeeshan Ahmad Khan, Hridip Kumar Sarma, and Asamanja Chattoraj. 2021. "Differential gene expression and immunohistochemical localization of the key melatonin biosynthesizing enzymes in the testis of zebrafish (*Danio rerio*)." *Biological Rhythm Research* 1-17. doi:10.1080/09291016.2021.1926078.

Doi, Masao, Jun Hirayama, and Paolo Sassone-Corsi. 2006. "Circadian regulator CLOCK is a histone acetyltransferase." *Cell* 125:497-508. doi:10.1016/j.cell.2006.03.033.

Duston, James, and Niall Bromage. 1986. "Photoperiodic mechanisms and rhythms of reproduction in the female rainbow trout." *Fish Physiology and Biochemistry* 2: 35-51. doi:10.1007/BF02264072.

Ebisawa, Takashi, Suresh Karne, Michael R. Lerner, and Steven M. Reppert. 1994. "Expression cloning of a high-affinity melatonin receptor from Xenopus dermal melanophores." *Proceedings of the National Academy of Sciences* 91:6133-6137. doi:10.1073/pnas.91.13.613.

Falcón, Jack. 1999. "Cellular circadian clocks in the pineal." *Progress in neurobiology* 58:121-162. doi:10.1016/S0301-0082(98)00078-1.

Falcón, Jacky, and Jean-Pierre Collin. 1991. "Pineal-retinal relationships: rhythmic biosynthesis and immunocytochemical localization of melatonin in the retina of the pike (*Esox lucius*)." *Cell and tissue research* 265:601-609. doi:10.1007/BF00340884.

Falcón, Jack, Y. Gothilf, S. L. Coon, G. Boeuf, and D. C. Klein. 2003. "Genetic, temporal and developmental differences between melatonin rhythm generating systems in the teleost fish pineal organ and retina." *Journal of neuroendocrinology* 15:378-382. doi:10.1046/j.1365-2826.2003.00993.x.

Falcón, Jack, Laurence Besseau, Sandrine Sauzet, and Gilles Boeuf. 2007. "Melatonin effects on the hypothalamo–pituitary axis in fish." *Trends in Endocrinology & Metabolism* 18:81-88. doi:10.1016/j.tem.2007.01.002.

Falcón, Jack, Herve Jose Migaud, Antonio Munoz-Cueto, and Manuel Carrillo. 2010. "Current knowledge on the melatonin system in teleost fish." *General and comparative endocrinology* 165:469-482. doi:10.1016/j.ygcen.2009.04.026.

Falcón, Jack, Laurence Besseau, Elodie Magnanou, Maria-Jesus Herrero, Masanori Nagai, and Gilles Boeuf. 2011. "Melatonin, the time keeper: biosynthesis and effects in fish." *Cybium* 35:3-18.

Falcón, Jack, Steven L. Coon, Laurence Besseau, Damien Cazaméa-Catalan, Michaël Fuentès, Elodie Magnanou, Charles-Hubert Paulin, Gilles Boeuf, Sandrine Sauzet, Even H Jørgensen, Sylvie Mazan, Yuri I Wolf, Eugene V Koonin, Peter J Steinbach, Susumu Hyodo, David C Klein. 2014. "Drastic neofunctionalization associated with evolution of the timezyme AANAT 500 Mya." *Proceedings of the National Academy of Sciences* 111:314-319. doi:10.1073/pnas.131263411.

Feldberg, Dov. 2021. "Biological Clock in Human Reproduction." In *Clinical Management of Infertility* 101-106. doi:10.1007/978-3-030-71838-1-7.

Félix, Francisca, Catarina CV Oliveira, and Elsa Cabrita. 2021. "Antioxidants in fish sperm and the potential role of melatonin." *Antioxidants* 10: 36. doi:10.3390/antiox10010036.

Foster, Russell G., and James Bellingham. 2004. "Inner retinal photoreceptors (IRPs) in mammals and teleost fish." *Photochemical & Photobiological Sciences* 3:617-627. doi:10.1039/B400092G.

Foster, Russell G., and Leon Kreitzman. 2005. "A Time to Reap." *BioScience* 55:795-797.

Gaeta, Laura Maria, Giulia Tozzi, Anna Pastore, Giorgio Federici, Enrico Bertini, and Fiorella Piemonte. 2002. "Determination of superoxide dismutase and glutathione peroxidase activities in blood of healthy pediatric subjects." *Clinica Chimica Acta* 322(1-2):117-120. doi:10.1016/S0009-8981(02)00163-8.

Galano, Annia, Dun Xian Tan, and Russel J. Reiter. 2013. "On the free radical scavenging activities of melatonin's metabolites, AFMK and AMK." *Journal of pineal research* 54:245-257. doi:10.1111/jpi.12010.

Gallo, Alessandra, Raffaele Boni, and Elisabetta Tosti. 2020. "Gamete quality in a multistressor environment." *Environment International* 138:105627. doi:10.1016/j.envint.2020.105627.

Gao, Yao, Chenhao Yang, Chencui Huang, Guoyun Zhang, Hong Ji, and Wuzi Dong. 2019. "The effects of melatonin supplement on paddlefish (*Polyodon spathula*) sperm quality and ATP content in sperm during in vitro storage." *Aquaculture* 503:475-482. doi: 10.1016/j.aquaculture.2019.01.010.

García-Allegue, R., J. A. Madrid, and F. J. Sánchez-Vázquez. 2001. "Melatonin rhythms in European sea bass plasma and eye: influence of seasonal photoperiod and water temperature." *Journal of pineal research* 31:68-75. doi:10.1034/j.1600-079X.2001.310110.x.

Giorgini, Elisabetta, Giorgia Gioacchini, Carla Conti, Paolo Ferraris, Simona Sabbatini, Giorgio Tosi, Chiara Carla Piccinetti, Lisa Vaccari, and Oliana Carnevali. 2012. "The role of melatonin on zebrafish follicle development: An FT-IR imaging approach." *Vibrational Spectroscopy* 62:279-285. doi:10.1016/j.vibspec.2012.06.009.

Gorman, Michael R. 2020. "Temporal organization of pineal melatonin signaling in mammals." *Molecular and Cellular Endocrinology* 503:110687. doi:10.1016/j.mce.2019.110687.

Goud, Anuradha P., Pravin T. Goud, Michael P. Diamond, Bernard Gonik, and Husam M. Abu-Soud. 2008. "Reactive Oxygen Species and Oocyte Aging: Role of Superoxide, Hydrogen Peroxide, and Hypochlorous Acid." *Free Radical Biology and Medicine* 44:1295-1304. doi:10.1016/j.freeradbiomed.2007.11.014.

Gozdowska, Magdalena, Ewa Sokołowska, Konrad Pomianowski, and Ewa Kulczykowska. 2022. "Melatonin and cortisol as components of the cutaneous stress response system in fish: Response to oxidative stress." *Comparative Biochemistry and Physiology Part A: Molecular & Integrative Physiology* 268:111207. doi:10.1016/j.cbpa.2022.111207.

Gupta, B. B. P. 2016. "Comparative endocrinology of the pineal organ: structural evolution, regulation and functions." In *Updates on integrative physiology and comparative endocrinology,* edited by Chandana Haldar, Sameer Gupta, and Soumik Goswami, 297-328. Varanasi: Publication Cell, Press and Publication Division, Banaras Hindu University.

Guraya, S. S., P. K. Sexena, and M. Gill. 1976. "Effect of long photoperiod on the maturation of ovary of the catfish, *Mystus tengara* (Ham.)." *Acta morphologica Neerlando-Scandinavica* 14:331-338.

Hafeez, Mehak, and Irfan Ahmad. 2021. "Melatonin and Seasonal Reproduction in Teleosts." In *Recent updates in molecular Endocrinology and Reproductive Physiology of Fish* 181-192. doi:10.1007/978-981-15-8369-8_13.

Hasan, Kazi Nurul, Mahammed Moniruzzaman, and Saumen Kumar Maitra. 2014. "Melatonin concentrations in relation to oxidative status and oocyte dynamics in the ovary during different reproductive phases of an annual cycle in carp *Catla catla*." *Theriogenology* 82:1173-1185. doi:10.1016/j.theriogenology.2014.08.001.

Hasan, Kazi Nurul, Palash Kumar Pal, and Saumen Kumar Maitra. 2016. "Diurnal and seasonal profiles of melatonin in the liver and their correlates of ovarian functions under natural photo-thermal conditions in adult carp *Catla catla*." *Biological Rhythm Research* 47:947-965. doi:10.1080/09291016.2016.1212535.

Hasan, Kazi Nurul, Palash Kumar Pal, and Saumen Kumar Maitra. 2020. "Temporal relationship between the levels of melatonin and different antioxidants in the liver of a surface feeding carp *Catla catla*." *Biological Rhythm Research* 51:373-391. doi:10.1080/09291016.2018.1533728.

Hazlerigg, D. G. 2001. "What is the role of melatonin within the anterior pituitary?." *The Journal of Endocrinology* 170:493-501. doi:10.1677/joe.0.1700493.

Hoffman, Roger A., and Russel J. Reiter. 1965. "Rapid pinealectomy in hamsters and other small rodents." *The Anatomical Record* 153:19-21. doi:10.1002/ar.1091530103.

Hong, Lu Yan, Wan Shu Hong, Wen Bo Zhu, Qiong Shi, Xin Xin You, and Shi Xi Chen. 2014. "Cloning and expression of melatonin receptors in the mudskipper *Boleophthalmus pectinirostris*: their role in synchronizing its semilunar spawning rhythm." *General and comparative endocrinology* 195:138-150. doi:10.1016/j.ygcen.2013.11.004.

Huang, Yu, Jia Li, Chao Bian, Ruihan Li, Xinxin You, and Qiong Shi. 2022. "Evolutionary Genomics Reveals Multiple Functions of Arylalkylamine N-acetyltransferase in Fish." *Frontiers in Genetics*: 1089. doi:10.3389/fgene.2022.820442.

Huether, Gerald. 1993. "The contribution of extrapineal sites of melatonin synthesis to circulating melatonin levels in higher vertebrates." *Experientia* 49:665-670. doi:10.1007/BF01923948.

Ianăş, O., R. Olinescu, and I. Badescu. 1991. "Melatonin Involvement in Oxidative Processes." *Revue Roumaine de Medecine - Serie Endocrinologie* 29(3-4):147-153.

Iigo, Masayuki, Kiyoshi Furukawa, Atsuhiko Hattori, Ritsuko Ohtani-Kaneko, Masayuki Hara, Takuro Suzuki, Mitsuo Tabata, and Katsumi Aida. 1997. "Ocular melatonin rhythms in the goldfish, Carassius auratus." *Journal of biological rhythms* 12:182-192.

Iigo, Masayuki, Teruo Azuma, and Munehico Iwata. 2007a. "Lack of circadian regulation of melatonin rhythms in the sockeye salmon (*Oncorhynchus nerka*) in vivo and in vitro." *Zoological science* 24:67-70. doi:10.2108/zsj.24.67.

Iigo, Masayuki, Tomotaka Abe, Saori Kambayashi, Kaoru Oikawa, Tomohiro Masuda, Kanta Mizusawa, Shoji Kitamura, Shoji Kitamura, Teruo Azuma, Yasuaki Takagi, Katsumi Aida, Tadashi Yanagisawa. 2007b. "Lack of circadian regulation of in vitro melatonin release from the pineal organ of salmonid teleosts." *General and comparative endocrinology* 154:91-97. doi:10.1016/j.ygcen.2007.06.013.

Isorna, Esther, María Aliaga-Guerrero, Abdeslam El M'Rabet, Arianna Servili, Jack Falcón, and José A. Muñoz-Cueto. 2011. "Identification of two arylalkylamine Nacetyltranferase 1 genes with different developmental expression profiles in the flatfish *Solea senegalensis.*" *Journal of Pineal Research* 51:434-444. doi:10.1111/j.1600-079X.2011.00907.x.

Itoh, Masanori T., Bunpei Ishizuka, Yasushi Kuribayashi, Akira Amemiya, and Yawara Sumi. 1999. "Melatonin, its precursors, and synthesizing enzyme activities in the human ovary." *Molecular human reproduction* 5:402-408. doi:10.1093/molehr/5.5.402.

Iuvone, P. Michael, and Joseph C. Besharse.1983. "Regulation of indoleamine N-acetyltransferase activity in the retina: effects of light and dark, protein synthesis inhibitors and cyclic nucleotide analogs." *Brain research* 273:111-119. doi:10.1016/0006-8993(83)91099-5.

Iuvone, P. Michael, Gianluca Tosini, Nikita Pozdeyev, Rashidul Haque, David C. Klein, and Shyam S. Chaurasia. 2005. "Circadian clocks, clock networks, arylalkylamine N-acetyltransferase, and melatonin in the retina." *Progress in retinal and eye research* 24:433-456. doi:10.1016/j.preteyeres.2005.01.003.

Jalabert, Bernard. 2005. "Particularities of reproduction and oogenesis in teleost fish compared to mammals." *Reproduction Nutrition Development* 45:261-279. doi:10.1051/rnd:2005019.

Jin, Ye Hwa, Jin Woo Park, Jung-Hyun Kim, and Joon Yeong Kwon. 2013. "The expression pattern of melatonin receptor 1a gene during early life stages in the Nile tilapia (*Oreochromis niloticus*)." *Development & Reproduction* 17:45. doi:10.12717/DR.2013.17.1.045.

Joy, K. P., and I. A. Khan. 1991. "Pineal-gonadal relationship in the teleost Channa punctatus (Bloch): evidence for possible involvement of hypothalamic serotonergic system." *Journal of pineal research* 11:12-22. doi:10.1111/j.1600-079X.1991.tb00821.x.

Kennaway, David J. 2005. "The role of circadian rhythmicity in reproduction." *Human reproduction update* 11:91-101. doi:10.1093/humupd/dmh054.

Khan, Izhar A., and Peter Thomas. 1996. "Melatonin influences gonadotropin II secretion in the Atlantic croaker (*Micropogonias undulatus*)." *General and Comparative Endocrinology* 104:231-242. doi:10.1006/gcen.1996.0166.

Khan, Zeeshan Ahmad, Thangal Yumnamcha, Chongtham Rajiv, Haobijam Sanjita Devi, Gopinath Mondal, Sh Dharmajyoti Devi, Rupjyoti Bharali, and Asamanja Chattoraj. 2016. "Melatonin biosynthesizing enzyme genes and clock genes in ovary and whole brain of zebrafish (*Danio rerio*): Differential expression and a possible interplay." *General and Comparative Endocrinology* 233:16-31. doi:10.1016/j.ygcen.2016.05.014.

Klein, David C. 2007. "Arylalkylamine N-acetyltransferase:"the Timezyme"." *Journal of biological chemistry* 282:4233-4237. doi:10.1074/jbc.R600036200.

Klosen, Paul, Sarawut Lapmanee, Carole Schuster, Beatrice Guardiola, David Hicks, Paul Pevet, and Marie Paule Felder-Schmittbuhl. 2019. "MT1 and MT2 melatonin receptors are expressed in nonoverlapping neuronal populations." *Journal of pineal research* 67:e12575. doi:10.1111/jpi.12575.

Konturek, P. C., T. Brzozowski, and S. J. Konturek, 2011. "Gut clock: implication of circadian rhythms in the gastrointestinal tract." *Journal of Physiology and Pharmacology* 62:139-150.

Kume, Kazuhiko, Mark J. Zylka, Sathyanarayanan Sriram, Lauren P. Shearman, David R. Weaver, Xiaowei Jin, Elizabeth S. Maywood, Michael H. Hastings, and Steven M. Reppert. 1999. "mCRY1 and mCRY2 are essential components of the negative limb of the circadian clock feedback loop." *Cell* 98:193-205. doi:10.1016/S0092-8674(00)81014-4.

Lerner, Aaron B., James D. Case, Yoshiyata Takahashi, Teh H. Lee, and Wataru Mori. 1958. "Isolation of melatonin, the pineal gland factor that lightens melanocytes." *Journal of the american chemical society* 80:2587-2587. doi:10.1021/ja01543a060.

Li, Ying, Guang Li, Haifang Wang, Jiulin Du, and Jun Yan. 2013. "Analysis of a gene regulatory cascade mediating circadian rhythm in zebrafish." *PLoS computational biology* 9:e1002940. doi:10.1371/journal.pcbi.1002940.

Lombardo, Francesco, Elisabetta Giorgini, Giorgia Gioacchini, Francesca Maradonna, Paolo Ferraris, and Oliana Carnevali. 2012. "Melatonin effects on *Fundulus heteroclitus* reproduction." *Reproduction, Fertility and Development* 24:794-803.

Lombardo, Francesco, Giorgia Gioacchini, Adele Fabbrocini, Michela Candelma, Raffaele D'Adamo, Elisabetta Giorgini, and Oliana Carnevali. 2014. "Melatonin-Mediated Effects on Killifish Reproductive Axis." *Comparative Biochemistry and Physiology - A Molecular and Integrative Physiology* 172:31-38. doi:10.1016/j.cbpa. 2014.02.008.

Lovenberg, W., Herbert Weissbach, and Sidney Udenfriend. 1962. "Aromatic LAmho acid decarboxylase." *J Biol Chem* 1:89-93.

Lovenberg, Walter, Eric Jequier, and Albert Sjoerdsma. 1967. "Tryptophan hydroxylation: measurement in pineal gland, brainstem, and carcinoid tumor." *Science* 155:217-219. doi: 10.1126/science.155.3759.21.

Lubzens, Esther, Graham Young, Julien Bobe, and Joan Cerdà. 2010. "Oogenesis in teleosts: how fish eggs are formed." *General and comparative endocrinology* 165:367-389. doi:10.1016/j.ygcen.2009.05.022.

Maitra, SK. 2011. Strategic management of photoperiodic schedule and melatonin profile in the regulation of carp reproduction. *In Carp: habitat, management and diseases*, edited by Sanders JD and Peterson SB, 1–20. New York: Nova Science Publishers, Inc.

Maitra, Saumen K., and Arun K. Ray. 2000. "Role of light in the mediation of acute effects of a single afternoon melatonin injection on steroidogenic activity of testis in the rat." *Journal of biosciences* 25:253-256. doi:10.1007/BF02703932.

Maitra, Saumen Kumar, and Kazi Nurul Hasan. 2016. "The role of melatonin as a hormone and an antioxidant in the control of fish reproduction." *Frontiers in endocrinology* 7:38. doi:10.3389/fendo.2016.00038.

Maitra, Saumen Kumar, Asamanja Chattoraj, and Sharmistha Bhattacharyya. 2005. "Implication of melatonin in oocyte maturation in Indian major carp *Catla catla*." *Fish Physiology and Biochemistry* 31: 201-207. doi:10.1007/s10695-006-0025-2.

Maitra, Saumen Kumar, Asamanja Chattoraj, Sourav Mukherjee, and Mahammed Moniruzzaman. 2013. "Melatonin: a potent candidate in the regulation of fish oocyte growth and maturation." *General and Comparative Endocrinology* 181:215-222. doi:10.1016/j.ygcen.2012.09.015.

Maitra, Saumen Kumar, Sourav Mukherjee, and Kazi Nurul Hasan. 2015. Melatonin: Endogenous sources and role in the regulation of fish reproduction Chapter 2. *In Indoleamines: sources, role in biological processes and health effects*, edited by Ángel Catalá, 43-77. New York: Nova Science Publishers.

Martinez-Chavez, Carlos Christian, Sammi Al-Khamees, Antonio Campos-Mendoza, David James Penman, and Herve Migaud. 2008. "Clock-Controlled Endogenous Melatonin Rhythms in Nile Tilapia (*Oreochromis niloticus niloticus*) and African Catfish (*Clarias gariepinus*)." *Chronobiology international* 25:31-49. doi:10.1080/07420520801917547.

Mattam, Ushodaya, and Anita Jagota. 2014. "Differential role of melatonin in restoration of age-induced alterations in daily rhythms of expression of various clock genes in suprachiasmatic nucleus of male Wistar rats." *Biogerontology* 15:257-268. doi:10.1007/s10522-014-9495-2.

McCord, Carey Pratt, and Floyd P Allen. 1917. "Evidences associating pineal gland function with alterations in pigmentation." *Journal of Experimental Zoology* 23: 207-224.

Menendez-Pelaez, Armando, Kimberly A. Howes, Aldo Gonzalez-Brito, and Russel J. Reiter. 1987."N-acetyltransferase activity, hydroxyindole-O-methyltransferase activity, and melatonin levels in the Harderian glands of the female Syrian hamster: changes during the light: dark cycle and the effect of 6-parachlorophenylalanine administration." *Biochemical and biophysical research communications* 145:1231-1238. doi:10.1016/0006-291X(87)91569-5.

Menendez-Pelaez, Armando, Burkhard Poeggeler, Russel J. Reiter, Lornell Barlow-Walden, Marta I. Pablos, and Dun-xian Tan. 1993. "Nuclear localization of melatonin in different mammalian tissues: immunocytochemical and radioimmunoassay evidence." *Journal of cellular biochemistry* 53:373-382. doi:10.1002/jcb.240530415.

Mhatre, Molina C., Albert S. van Jaarsveld, and Russel J. Reiter. 1988. "Melatonin in the lacrimal gland: first demonstration and experimental manipulation." *Biochemical and biophysical research communications* 153:1186-1192. doi:10.1016/S0006-291X(88)81353-6.

Migaud, Herve, Andrew Davie, and J. F. Taylor. 2010. "Current knowledge on the photoneuroendocrine regulation of reproduction in temperate fish species." *Journal of Fish Biology* 76:27-68. doi:10.1111/j.1095-8649.2009.02500.x.

Mondal, Pradip, Kazi Nurul Hasan, Palash Kumar Pal, and Saumen Kumar Maitra. 2017. "Influences of exogenous melatonin on the oocyte growth and oxidative status of ovary during different reproductive phases of an annual cycle in carp *Catla catla*." *Theriogenology* 87:349-359. doi:10.1016/j.theriogenology.2016.09.021.

Moniruzzaman, Mahammed, and Saumen Kumar Maitra. 2012. "Influence of altered photoperiods on serum melatonin and its receptors (MT1 and MT2) in the brain, retina, and ovary in carp *Catla catla*." *Chronobiology International* 29:175-188. doi:10.3109/07420528.2011.645753.

Moniruzzaman, Mahammed, Kazi Nurul Hasan, and Saumen Kumar Maitra. 2016. "Melatonin actions on ovaprim (synthetic GnRH and domperidone)-induced oocyte maturation in carp." *Reproduction* 15: 285-296. doi:10.1530/REP-15-0391.

Mukherjee, Sourav, Mahammed Moniruzzaman, and Saumen Kumar Maitra. 2014a. "Daily and seasonal profiles of gut melatonin and their temporal relationship with pineal and serum melatonin in carp *Catla catla* under natural photo-thermal conditions." *Biological Rhythm Research* 45:301-315. doi:10.1080/09291016.2013.817139.

Mukherjee, Sourav, Mahammed Moniruzzaman, and Saumen Kumar Maitra. 2014b. "Impact of artificial lighting conditions on the diurnal profiles of gut melatonin in a surface dwelling carp (*Catla catla*)." *Biological Rhythm Research* 45:831-848. doi:10.1080/09291016.2014.923618.

Nagahama, Y. 1994. "Endocrine regulation of gametogenesis in fish." *International Journal of Developmental Biology 38*:217–229.

Nagahama, Yoshitaka. 1997. "17α, 20β-Dihydroxy-4-pregnen-3-one, a maturation-inducing hormone in fish oocytes: mechanisms of synthesis and action." *Steroids* 62:190-196. doi:10.1016/S0039-128X(96)00180-8.

Nagahama, Yoshitaka, and Masakane Yamashita. 2008. "Regulation of oocyte maturation in fish." *Development, growth & differentiation* 50:S195-S219 doi:10.1111/j.1440-169X.2008.01019.x.

Nakamura, Yasuhiko, Hiroshi Tamura, Hisako Takayama, and Hiroshi Kato. 2003. "Increased endogenous level of melatonin in preovulatory human follicles does not directly influence progesterone production." *Fertility and sterility* 80:1012-1016. doi:10.1016/S0015-0282(03)01008-2.

Nowak, Jerzy Z., Jolanta B. Zawilska, and Agata Woldan-Tambor. 1995. "Arylamine and arylalkylamine N-acetyltransferases in retina, pineal gland, brain and liver of chicks: a comparative study." *Neurochemistry international* 27:285-297. doi:10.1016/0197-0186(95)00031-3.

Ogiwara, Katsueki, and Takayuki Takahashi. 2016. "A dual role for melatonin in medaka ovulation: Ensuring prostaglandin synthesis and actin cytoskeleton rearrangement in follicular cells." *Biology of Reproduction* 94:64-1. doi:10.1095/biolreprod.115.133827.

Olcese, James M. 2020. "Melatonin and Female Reproduction: An Expanding Universe." *Frontiers in Endocrinology* 11:85. doi:10.3389/fendo.2020.00085.

Pal, Palash Kumar, and Saumen Kumar Maitra. 2018. "Response of gastrointestinal melatonin, antioxidants, and digestive enzymes to altered feeding conditions in carp (*Catla catla*)." *Fish physiology and biochemistry* 44:1061-1073. doi:10.1007/s10695-018-0494-0.

Pal, Palash Kumar, Kazi Nurul Hasan, and Saumen Kumar Maitra. 2012. "Functional Pleiotropy of Melatonin in the Regulation of Reproduction: An Overview." *Journal of Endocrinology and Reproduction* 16:33-43.

Pal, Palash Kumar, Kazi Nurul Hasan, and Saumen Kumar Maitra. 2016a. "Temporal relationship between the daily profiles of gut melatonin, oxidative status and major digestive enzymes in carp *Catla catla*." *Biological Rhythm Research* 47:755-771. doi:10.1080/09291016.2016.1191697.

Pal, Palash Kumar, Kazi Nurul Hasan, and Saumen Kumar Maitra. 2016b. "Gut melatonin response to microbial infection in carp *Catla catla*." *Fish physiology and biochemistry* 42:579-592. doi:10.1007/s10695-015-0161-7.

Pan, Xiaoyue, Meredith J. Taylor, Emma Cohen, Nazeeh Hanna, and Samantha Mota. 2020. "Circadian clock, time-restricted feeding and reproduction." *International journal of molecular sciences* 21:831. doi:10.3390/ijms21030831.

Patiño, Reynaldo, and Peter Thomas. 1990. "Effects of gonadotropin on ovarian intrafollicular processes during the development of oocyte maturational competence in a teleost, the Atlantic croaker: evidence for two distinct stages of gonadotropic control of final oocyte maturation." *Biology of Reproduction* 43:818-827. doi:10.1095/biolreprod43.5.818.

Patiño, Reynaldo, Goro Yoshizaki, Peter Thomas, and Hirohiko Kagawa. 2001. "Gonadotropic control of ovarian follicle maturation: the two-stage concept and its mechanisms." *Comparative Biochemistry and Physiology Part B: Biochemistry and Molecular Biology* 129:427-439. doi: 10.1016/S1096-4959(01)00344-X.

Paulin, Charles-Hubert, Damien Cazaméa-Catalan, Bina Zilberman-Peled, Patricia Herrera-Perez, Sandrine Sauzet, Elodie Magnanou, Michael Fuentès, Yoav Gothilf, Jose Antonio Muñoz-Cueto, Jack Falcón, Laurence Besseau. 2015. "Subfunctionalization of arylalkylamine N-acetyltransferases in the sea bass *Dicentrarchus labrax*: two-ones for one two." *Journal of Pineal Research* 59:354-364. doi:10.1111/jpi.12266.

Peuhkuri, Katri, Nora Sihvola, and Riitta Korpela. 2012. "Dietary factors and fluctuating levels of melatonin." *Food & nutrition research* 56:17252. doi:10.3402/fnr.v56i0.17252.

Pevet, P. 2000. "Melatonin and biological rhythms." *Neurosignals* 9:203-212. doi:10.1159/000014640.

Pevet, Paul, Laurence Agez, B. Bothorel, Michel Saboureau, François Gauer, Virginie Laurent, and Mireille Masson-Pévet. 2006. "Melatonin in the multi-oscillatory mammalian circadian world." *Chronobiology international* 23: 39-51. doi:10.1080/07420520500482074.

Pfeffer, Martina, Horst-Werner Korf, and Helmut Wicht. 2017. "The role of the melatoninergic system in light-entrained behavior of mice." *International Journal of Molecular Sciences* 18:530. doi:10.3390/ijms18030530.

Popek, Włodzimierz, Piotr Epler, Krzysztof Bieniarz, and Mirosława Sokołowska-Mikołajczyk. 1997. "Contribution of factors regulating melatonin release from pineal gland of carp (*Cyprinus carpio* L.) in normal and in polluted environments." *Fisheries & Aquatic Life* 5:59-75.

Porter, M. J. R., S. O. Stefansson, G. Nyhammer, Ø. Karlsen, B. Norberg, and N. R. Bromage. 2000. "Environmental influences on melatonin secretion in Atlantic cod (*Gadus morhua* L.) and their relevance to commercial culture." *Fish Physiology and Biochemistry* 23:191-200. doi:10.1023/A:1007816426509.

Rad, Ferit, Sezai Bozaoğlu, Serap Ergene Gözükara, Arzu Karahan, and Gülderen Kurt. 2006. "Effects of different long-day photoperiods on somatic growth and gonadal development in Nile tilapia (*Oreochromis niloticus* L.)." *Aquaculture* 255:292-300. doi:10.1016/j.aquaculture.2005.11.028.

Raikhlin, N. T., and I. M. Kvetnoy. I. M. 1974. "Lightening effect of the extract of human appendix mucosa on frog skin melanophores." *Bulletin of Experimental Biology and Medicine* 8:114-116.

Raikhlin, Nathan T., and Igor M. Kvetnoy. 1976. "Melatonin and enterochromaffine cells." *Acta Histochemica* 55:19-24. doi:10.1016/S0065-1281(76)80092-X.

Reiter, Russel J. 1991. "Pineal melatonin: cell biology of its synthesis and of its physiological interactions." *Endocrine reviews* 12:151-180. doi:10.1210/edrv-12-2-151.

Reiter, Russel J., Sergio A. Rosales-Corral, Lucien C. Manchester, and Dun-Xian Tan. 2013. "Peripheral reproductive organ health and melatonin: ready for prime time." *International journal of molecular sciences* 14:7231-7272. doi:10.3390/ijms14047231.

Richter, Hans G., Claudia Torres-Farfan, Pedro P. Rojas-García, Carmen Campino, Fernando Torrealba, and María Serón-Ferré. 2004. "The circadian timing system: making sense of day/night gene expression." *Biological Research* 37:11-28. doi:10.4067/S0716-97602004000100003.

Rodriguez, Carmen, Juan C. Mayo, Rosa M. Sainz, Isaac Antolín, Federico Herrera, Vanesa Martín, and Russel J. Reiter. 2004. "Regulation of Antioxidant Enzymes: A Significant Role for Melatonin." *Journal of Pineal Research* 36:1-9. doi:10.1046/j.1600-079X.2003.00092.x.

Ronnberg, Lars, Antti Kauppila, Juhani Leppaluoto, Hannu Martikainen, and Olli Vakkuri. 1990. "Circadian and seasonal variation in human preovulatory follicular fluid melatonin concentration." *The Journal of Clinical Endocrinology & Metabolism* 71:493-496. doi:10.1210/jcem-71-2-493.

Saha, Saurav, Kshetrimayum Manisana Singh, and Braj Bansh Prasad Gupta. 2019. "Melatonin synthesis and clock gene regulation in the pineal organ of teleost fish compared to mammals: Similarities and differences." *General and comparative Endocrinology* 279:27-34. doi:10.1016/j.ygcen.2018.07.010.

Salhab, Mohamad, Sophie Dhorne-Pollet, Sylvain Auclair, Catherine Guyader-Joly, Daphné Brisard, Rozenn Dalbies-Tran, Joelle Dupont, Claire Ponsart, Pascal Mermillod, and Svetlana Uzbekova. 2013. "*In vitro* maturation of oocytes alters gene expression and signaling pathways in bovine cumulus cells." *Molecular reproduction and development* 80:166-182. doi:10.1002/mrd.22148.

Sánchez-Vázquez, Francisco Javier, Jose Fernando López-Olmeda, Luisa Maria Vera, Herve Migaud, Marcos Antonio López-Patiño, and Jesús M. Míguez. 2019. "Environmental cycles, melatonin, and circadian control of stress response in fish." *Frontiers in endocrinology* 10:279. doi:10.3389/fendo.2019.00279.

Saxena, P. K., and K. Anand. 1977. "A comparison of ovarian recrudescence in the catfish, *Mystus tengara* (Ham.), exposed to short photoperiods, to long photoperiods, and to melatonin." *General and Comparative Endocrinology* 33:506-511.

Sellix, Michael T. 2015. "Circadian clock function in the mammalian ovary." *Journal of biological rhythms* 30:7-19. doi:10.1177/0748730414554222.

Servili, Arianna, Patricia Herrera-Pérez, María del Carmen Rendón, and José Antonio Muñoz-Cueto. 2013. "Melatonin Inhibits GnRH-1, GnRH-3 and GnRH Receptor Expression in the Brain of the European Sea Bass, *Dicentrarchus labrax*."

International Journal of Molecular Sciences 14:7603-7616. doi:10.3390/ijms14047603.

Servili, Arianna, Adelino VM Canario, Olivier Mouchel, and José Antonio Muñoz-Cueto. 2020. "Climate change impacts on fish reproduction are mediated at multiple levels of the brain-pituitary-gonad axis." *General and Comparative Endocrinology* 291:113439. doi:10.1016/j.ygcen.2020.113439.

Seth, Mohua, and Saumen Kumar Maitra. 2010. "Photoreceptor proteins and melatonin rhythm generating AANAT in the carp pineal: temporal organization and correlation with natural photo-thermal cues." *Journal of Photochemistry and Photobiology B: Biology* 99:21-28. doi:10.1016/j.jphotobiol.2010.01.010.

Shi, Qiong, Hironori Ando, Steven L. Coon, Shunpei Sato, Masatoshi Ban, and Akihisa Urano. 2004. "Embryonic and post-embryonic expression of arylalkylamine N-acetyltransferase and melatonin receptor genes in the eye and brain of chum salmon (*Oncorhynchus keta*)." *General and comparative endocrinology* 136:311-321. doi:10.1016/j.ygcen.2004.01.004.

Shin, Hyun Suk, Jehee Lee, and Cheol Young Choi. "Effects of LED light spectra on oxidative stress and the protective role of melatonin in relation to the daily rhythm of the yellowtail clownfish, *Amphiprion clarkii*. 2011. " *Comparative Biochemistry and Physiology Part A: Molecular & Integrative Physiology* 160:221-228. doi:10.1016/j.cbpa.2011.06.002.

Singh, T. P., and P. Lal, 1994. "Endocrine physiology of reproduction in Indian catfish," *In Advances in Fish Biology*, edited by H. R., 147–154. Delhi: Singh Hindustan Publishing Corporation.

Srivastava, S. J., and R. Singh. 1992. "Effects of constant photoperiod-temperature regimes on the ovary and pituitary gonadotropic cells of the murrel *Channa punctatus* (Bloch) during the annual reproductive cycle." *Biological structures and morphogenesis* 4:28-31.

Srivastava, S. J., and R. Singh. 1993. "Effects of constant photoperiod-temperature regimes on testes and pituitary gonadotropic cells of the murrel, *Channa punctatus* (Bloch) during the annual reproductive cycle." In *Annales d'endocrinologie*, 203-206.

Sundararaj, B. I. 1981. Aquaculture development and coordination programme. Reproductive physiology of teleost fishes. A review of present knowledge and needs for future research. Food and Agriculture Organization of the United Nations, Rome (Italy).

Sundararaj, Bangalore I., and P. Keshavanath. 1976. "Effects of melatonin and prolactin treatment on the hypophysial-ovarian system in the catfish, *Heteropneustes fossilis* (Bloch)." *General and comparative endocrinology* 29:84-96. doi:10.1016/0016-6480(76)90010-1 doi:10.1016/0016-6480(76)90010-1.

Suresh, D. V. N. S., V. V. Baile, and PD Prasada Rao. 2008. "Annual reproductive phase-related profile of sex steroids and their carrier, SHBG, in the Indian major carp, *Labeo rohita*." *General and comparative endocrinology* 159:143-149. doi:10.1016/j.ygcen.2008.08.004.

Swanson, L. W., W. M. Cowan, and E. G. Jones. 1974. "An autoradiographic study of the efferent connections of the ventral lateral geniculate nucleus in the albino rat and the cat." *Journal of Comparative Neurology* 156:143-163. doi:10.1002/cne.901560203

Takahashi, Takayuki, and Katsueki Ogiwara. 2021. "Roles of melatonin in the teleost ovary: A review of the current status." *Comparative Biochemistry and Physiology Part A: Molecular & Integrative Physiology* 254:110907. doi:10.1016/j.cbpa. 2021.110907.

Tamura, Hiroshi, Yasuhiko Nakamura, Ahmet Korkmaz, Lucien C. Manchester, Dun-Xian Tan, Norihiro Sugino, and Russel J. Reiter. 2009. "Melatonin and the ovary: physiological and pathophysiological implications." *Fertility and sterility* 92:328-343. doi:10.1016/j.fertnstert.2008.05.016.

Tamura, Hiroshi, Mai Jozaki, Manabu Tanabe, Yuichiro Shirafuta, Yumiko Mihara, Masahiro Shinagawa, Isao Tamura, Ryo Maekawa, Shun Sato, Toshiaki Taketani, Akihisa Takasaki, Russel J Reiter, Norihiro Sugino. 2020. "Importance of melatonin in assisted reproductive technology and ovarian aging." *International journal of molecular sciences* 21:1135. doi:10.3390/ijms21031135.

Tan, Dun-Xian. 1993. "Melatonin: a potent, endogenous hydroxyl radical scavenger." *Endocrine Journal* 1: 57-60.

Tan, Dun-Xian, Xiaodong Zheng, Jin Kong, Lucien C. Manchester, Ruediger Hardeland, Seok Joong Kim, Xiaoying Xu, and Russel J. Reiter. 2014. "Fundamental issues related to the origin of melatonin and melatonin isomers during evolution: relation to their biological functions." *International Journal of Molecular Sciences* 15:15858-15890. doi:10.3390/ijms150915858.

Tosini, Gianluca, and Michael Menaker. 1996. "Circadian rhythms in cultured mammalian retina." *Science* 272:419-421. doi:10.1126/science.272.5260.419.

Tosini, Gianluca, and Michael Menaker. 1998. "The clock in the mouse retina: melatonin synthesis and photoreceptor degeneration." *Brain research* 789:221-228. doi:10.1016/S0006-8993(97)01446-7.

Tosini, Gianluca, Kenkichi Baba, Christopher K. Hwang, and P. Michael Iuvone. 2012. "Melatonin: an underappreciated player in retinal physiology and pathophysiology." *Experimental eye research* 103:82-89. doi:0.1016/j.exer.2012.08.009.

Trippel, Edward A., and Steven RE Neil. 2003. "Effects of photoperiod and light intensity on growth and activity of juvenile haddock (*Melanogrammus aeglefinus*)." *Aquaculture* 217:633-645. doi:10.1016/S0044-8486(02)00198-9.

Tsutsui, Kazuyoshi. 2009. "A new key neurohormone controlling reproduction, gonadotropin-inhibitory hormone (GnIH): Biosynthesis, mode of action and functional significance." *Progress in neurobiology* 88:76-88. doi:10.1016/ j.pneurobio.2009.02.003.

Tsutsui, Kazuyoshi, and Takayoshi Ubuka. 2020. "Discovery of gonadotropin-inhibitory hormone (GnIH), progress in GnIH research on reproductive physiology and behavior and perspective of GnIH research on neuroendocrine regulation of reproduction." *Molecular and Cellular Endocrinology* 514:110914. doi:10.1016/ j.mce.2020.110914.

Tsutsui, Kazuyoshi, Takayoshi Ubuka, George E. Bentley, and Lance J. Kriegsfeld. 2012. "Gonadotropin-inhibitory hormone (GnIH): discovery, progress and prospect." *General and comparative endocrinology* 177:305-314. doi:10.1016/j.ygcen.2012. 02.013.

Türker, Ali, MURAT YİĞİT, and Sebahattin Ergün. 2005. "Growth and feed utilization in juvenile Black Sea turbot (*Psetta maeotica*) under different photoperiod regimes." *Turkish Journal of Veterinary and Animal Sciences* 29:1203-1208.

Ubuka, Takayoshi, George E. Bentley, Kazuyoshi Ukena, John C. Wingfield, and Kazuyoshi Tsutsui. 2005. "Melatonin Induces the Expression of Gonadotropin-Inhibitory Hormone in the Avian Brain." *Proceedings of the National Academy of Sciences of the United States of America* 120:3052-3057. doi:10.1073/pnas.0403840102.

Vakkuri, Olli, Hannu Rintamäki, and Juhani Leppäluoto.1985. "Presence of immunoreactive melatonin in different tissues of the pigeon (*Columba livia*)." *General and comparative endocrinology* 58:69-75. doi:10.1016/0016-6480(85)90136-4.

Valenzuela, F. J., J. Vera, C. Venegas, F. Pino, and C. Lagunas. 2015. "Circadian System and Melatonin Hormone: Risk Factors for Complications during Pregnancy." *Obstetrics and Gynecology International* 2015:825802. doi:10.1155/2015/825802.

Velarde, Elena, Jose Miguel Cerdá-Reverter, Angel Luis Alonso-Gómez, Elisa Sánchez, Esther Isorna, and María Jesús Delgado. 2010. "Melatonin-synthesizing enzymes in pineal, retina, liver, and gut of the goldfish (*Carassius*): mRNA expression pattern and regulation of daily rhythms by lighting conditions." *Chronobi ology International* 27:1178-1201. doi:10.3109/07420528.2010.496911.

Verghese, P. U. 1976. "Internal rhythm of sexual cycle in a carp *Cirrhina reba* (Ham.) under artificial conditions of darkness [India]." *Journal of the Inland Fisheries Society of India*, 182-188.

Voisin, Pierre, M. A. Namboodiri, and D. C. Klein. 1984. "Arylamine N-acetyltransferase and arylalkylamine N-acetyltransferase in the mammalian pineal gland." *Journal of Biological Chemistry* 259:10913-10918. doi:10.1016/S0021-9258(18)90600-9.

Vollrath, Lutz. 1981. "*The pineal organ.*" In Handbuch der mikroskopischen Anatomie des Menschen Handbook of Mikroscopic, Springer Berlin, Heidelberg.

von Gall, Charlotte, Martine L. Garabette, Christian A. Kell, Sascha Frenzel, Faramarz Dehghani, Petra-Maria Schumm-Draeger, David R. Weaver, Horst-Werner Korf, Michael H. Hastings, and Jörg H. Stehle. 2002. "Rhythmic gene expression in pituitary depends on heterologous sensitization by the neurohormone melatonin." *Nature neuroscience* 5:234-238. doi:10.1038/nn806.

Vriend, Jerry, and Russel J. Reiter. 2014. "Melatonin as a proteasome inhibitor. Is there any clinical evidence?" *Life sciences* 115:8-14. doi:10.1016/j.lfs.2014.08.024.

Vriend, Jerry, and Russel J. Reiter. 2015. "Melatonin feedback on clock genes: a theory involving the proteasome." *Journal of pineal research* 58:1-11. doi:10.1111/jpi.12189.

Vuilleumier, Robin, Gilles Boeuf, Michael Fuentes, Walter J. Gehring, and Jack Falcón. 2007. "Cloning and early expression pattern of two melatonin biosynthesis enzymes in the turbot (*Scophthalmus maximus*)." *European Journal of Neuroscience* 25:3047-3057. doi:10.1111/j.1460-9568.2007.05578.x.

Zelko, Igor N., Thomas J. Mariani, and Rodney J. Folz. 2002. "Superoxide Dismutase Multigene Family: A Comparison of the CuZn-SOD (SOD1), Mn-SOD (SOD2), and EC-SOD (SOD3) Gene Structures, Evolution, and Expression." *Free Radical Biology and Medicine* 33:337-349. doi:10.1016/S0891-5849(02)00905-X.

Zhang, W. L., Z. L. Zhang, J. Peng, S. T. Yang, J. X. Chen, C. X. Wang, and D. W. Tong. 2021. "Expression of arylalkylamine n-acetyltransferase (AANAT) and acetylserotonin o-methyltransferase (ASMT) in the corpus luteum of pregnant sows and synthesis of melatonin in luteal cells." *Cell and tissue research*, 1-13. doi:10.1007/s00441-021-03556-y.

Zohar, Yonathan, José Antonio Muñoz-Cueto, Abigail Elizur, and Olivier Kah. 2010. "Neuroendocrinology of reproduction in teleost fish." *General and comparative endocrinology* 165:438-455. doi:10.1016/j.ygcen.2009.04.017.

Chapter 3

Biotic Interaction Gradients between Native and Invasive Aquatic Gastropods

Kiza K. Gates[1,2,*] and Billie L. Kerans[1]

[1]Montana State University, Bozeman, MT, USA

[2]Washington Department of Fish and Wildlife, WA, USA

Abstract

Much research has been devoted to understanding how interactions shift along abiotic gradients of stress, but less is known about biotic gradients and how they can affect the strength and intensity of interactions between species. We explored the interaction between the native aquatic gastropod, *Valvata humeralis,* and invasive aquatic gastropod, *Potamopyrgus antipodarum,* along a biotic gradient of density in a system where we suspected that an interaction shift was occurring. We experimentally quantified the intra- and interspecific net effects of these species on each other and tested the results in a field setting by comparing them with field surveys. *Valvata humeralis* growth was significantly greater in treatments containing *V. humeralis* and *P. antipodarum* at $400/m^2$ each (total gastropod density $800/m^2$) when compared to growth of *V. humeralis* alone at $400/m^2$. The facilitative growth effect disappeared at higher snail densities indicating it was density specific. Field surveys indicated that competition between the species at medium densities did not negatively affect *V. humeralis* growth, but did not show the same facilitation seen in the experiment. Experimental and field results combined suggest that the shifting interaction between species is intertwined with other biotic and abiotic factors in the field. Our results suggest that interactions between a native and an invasive species can

* Corresponding Author's Email: kizagates@gmail.com.

In: Advances in Animal Science and Zoology. Volume 20
Editor: Owen P. Jenkins
ISBN: 979-8-88697-199-6

shift along a biotic density gradient potentially influencing invasive species impact.

Introduction

Interactions between neighboring species are often bidirectional existing along a gradient from competition to facilitation, but are rarely studied as such (Bronstein 1994; Bruno et al. 2003). Bidirectional interactions are the result of changes in the interaction components that sum to make the net interaction. Interactions change direction along environmental stress gradients (Bertness and Callaway 1994; Callaway and Walker 1997; Brooker and Callaghan 1998; Kawai and Tokeshi 2007; Brooker et al. 2008) over different spatial scales (Bertness et al. 1999; Callaway et al. 2002) and through time (Tielborger and Kadmon 2000; Kawai and Tokeshi 2007). These bidirectional interactions create non-linear interaction coefficients which produce unexpected outcomes between species that are thought to compete (Hernandez 1998; Dick et al. 2014). More exploration of the feedbacks that create bidirectional interactions between species is needed to understand ecological dynamics (Agrawal et al. 2007).

Much of our knowledge of changing interactions comes from research exploring the Stress-Gradient Hypothesis (SGH) with sessile organisms and abiotic stress gradients. The SGH predicts that in high stress environments, the amelioration of stress for one species, the beneficiary, by another species, the benefactor, produces a facilitation that offsets competition for resources. In other words, the occurrence of competitive and facilitative interactions varies inversely with increasing stress (Bertness and Callaway 1994; Maestre et al. 2009). Facilitation typically takes the form of habitat enhancement where habitat quality is increased (through resources or non-resources) for one species by the presence of another (Maestre et al. 2009). This form of facilitation has been commonly identified among plants (Went 1942; Callaway 1995; Callaway and Walker 1997) and intertidal marine organisms (Bertness 1989; Bertness et al. 1999; Kawai and Tokeshi 2004). Stresses that can be ameliorated by the presence of another species include exposure to heat (Kawai and Tokeshi 2006; Kawai and Tokeshi 2007), wave-action (Kawai and Tokeshi 2006; Kawai and Tokeshi 2007), or anthropogenic disturbance (Villarreal-Barajas and Martorell 2009). In all of these cases, the facilitation and shifting interaction only become apparent in high stress environments when the presence of the benefactor is facilitative for the beneficiary outweighing the competitive interactions under lower stress. In this manner, the abiotic gradient of stress determines the net species interaction.

Biotic stress gradients have not been extensively explored despite their potential to create bidirectional interactions that influence species coexistence similar to abiotic gradients. Some biotic gradient research has been conducted exploring plant interactions along a consumer stress gradient of grazing intensity (Smit et al. 2007; Smit et al. 2009; Graff et al. 2007; Graff and Aguiar 2011). This work has found evidence of interaction shifts along a consumer stress gradient, but results have been mixed with some evidence aligning with the SGH (Graff and Aguiar 2011) and some requiring modifications of the SGH (Smit et al. 2009; Soliveres et al. 2011). Another potential biotic stress gradient that has received even less attention is benefactor density (Bruno and Bertness 2001; Bulleri et al. 2011). Preliminary research with benefactor density gradients indicates that the interaction between benefactor and beneficiary can change along a benefactor density gradient, but that the outcome may also be dependent on abiotic stress levels (Irving and Bertness 2009) or individual species traits (Bulleri et al. 2011). More research is needed to better understand the influence of benefactor density gradients on species coexistence.

Inconsistent outcomes from the interactions of biologically similar species often occur in invasion ecology. Invasive species, non-native species with a history of causing ecological or economic impact, often only exhibit invasive behavior in particular geographic locations (Kolar and Lodge 2001) and many introductions do not appreciably change the receiving community (Williamson 1996; Davis 2003). Failure of an invasive species to impact native species can sometimes be attributed to their inability to proceed through the entire invasion process; yet, there can still be considerable variation in native species response among invasion sites. The mechanism of this variation is uncertain. The potential influence of positive interactions between invasive and native species is rarely considered despite the prevalence of positive interactions (Bruno et al. 2005) and existing evidence that interactions may shift along invasive density gradients promoting coexistence with native species (Straube et al. 2009).

Previous research with the invasive aquatic gastropod *Potamopyrgus antipodarum* is suggestive of shifting interactions with native invertebrates. *Potamopyrgus antipodarum* is known for achieving densities more than 250,000 snails/m^2 (Richards et al. 2001; Kerans et al. 2005; Alonso and Castro-díez 2008). Factors which have been suggested to contribute to the success of *P. antipodarum* in introduced systems include their parthenogenic and ovoviviparous reproduction (Dorgelo 1988), high secondary production (Hall et al. 2006), ability to reproduce throughout most of the year (Dorgelo 1987), and ability to quickly re-colonize disturbed areas (Schreiber et al. 2003). Yet while *P. antipodarum* has invaded many Western waters, there is considerable variability in invasion establishment among systems (Geist et al. 2021). *Potamopyrgus antipodarum* can dominate invertebrate numbers and biomass in established areas (Kerans et al. 2005; Hall et al. 2003; Geist et al. 2021) sometimes with the aid of native species (Riley et al. 2008), but also experiences extreme population crashes (Dorgelo 1988) and coexist through time with native gastropod species (Siegismund and Hylleberg 1987). For example, in the Vista reach of the Middle Snake River high densities of the native aquatic gastropod *Valvata humeralis* coexist with *P. antipodarum* (Gates 2012). Despite suspected competition with *P. antipodarum* (Lysne and Koetsier 2008); the native *V. humeralis* has maintained large populations with *P. antipodarum* for over a decade in some locations. We hypothesized that coexistence of these two species was promoted by an interaction that shifts from negative to positive along a benefactor density gradient where *P. ant*ipodarum is the benefactor and *V. humeralis* is the beneficiary.

We explored the interaction between a native species (beneficiary) and invasive species (benefactor) along a biotic stress gradient of benefactor density in a system where we suspected that an interaction shift was occurring. We experimentally quantified the intra- and interspecific net effects of the native species *V. humeralis* and the invasive species *P. antipodarum* along a gradient of density and species proportion. We tested the hypothesis that the effect of an invasive species on a native species changes with density of the invasive species using a field experiment and tested the results in a field setting by comparing them with field surveys.

Methods

Study Species and Site

Research was conducted in the Vista reach of the Middle Snake River below American Falls Dam, U.S.A. (Latitude 42°46'03," Longitude 112°52'46," RM 713). Previous surveys indicated that *V. humeralis* and *P. antipodarum* have coexisted in this reach for at least 15 years (R. Newman, Bureau of Reclamation, personal communication). The Vista reach is characterized by a narrow channel allowing most of the riverbed to remain watered throughout the year despite dramatic changes in discharge from American Falls dam upriver (11-425 m^3/s 2000-2010 mean low and high discharge). Water temperature fluctuates from 5-22°C throughout the year.

Our target native species, *V. humeralis,* is a colonial species that forms distinctive populations that can be found from year to year in the same location and are known to coexist with *P. antipodarum* in the Middle Snake River making them a good candidate for interaction studies (R. Newman, personal communication). In addition, *V. humeralis* is semelparous reproducing in late summer in the Snake River allowing a cohort to be followed throughout the year. Our target invasive species, *P. antipodarum*, is a global invader that has been successful in lentic and lotic freshwaters of Australia (Schrieber et al. 2003), Europe (Dorgelo 1987), and more recently North America (Bowler 1991; Langenstein and Bowler 1991; Bowler and Frest 1992; Zaranko et al. 1997; Richards et al. 2001) among other regions (Geist et al. 2021). A broad diet and physiological tolerance (Winterbourn 1970, Dorgelo et al. 1995; Collier and Winterbourn 2000) and continuous reproduction give *P. antipodarum* a potentially competitive advantage in many habitats (Loo et al. 2007). Other gastropod species within the Vista reach include the native species *Gyraulus parvus*, *Fluminicola fuscus*, *Lymnaidae* sp., *Physella gyrina, Stagnicola* sp., *Valvata utahensis*, and *Vorticifex effusa*.

Field Experiment

We assessed the effects of intraspecific and interspecific density on the growth and mortality of *V. humeralis* and *P. antipodarum* with a field cage experiment modified from Underwood (1997) and Cope and Winterbourn (2004). We used growth and mortality as indicators of interaction direction and intensity. There were three densities levels: low (6 snails/cage or 400 snails/m^2), medium (12 snails/cage or 800 snails/m^2), and high (18 snails/cage or 1200 snails/m^2) (Table 1). The low-density treatment corresponded with the average *V. humeralis* density found in the field during population surveys in 2008 (Gates 2012). We created seven treatments including low density *V. humeralis,* medium density *V. humeralis,* low density *P. antipodarum,* medium density *P. antipodarum,* medium density *V. humeralis* and *P. antipodarum* (1:1), high density *V. humeralis* and *P. antipodarum* (1:2), and high-density *V. humeralis* and *P. antipodarum* (2:1). Within these treatments, we measured growth and mortality of *V. humeralis* in all five treatments containing *V. humeralis* and we measured growth and mortality of *P. antipodarum* in all five treatments containing *P. antipodarum* creating 10 levels of the response variable. We replicated treatments three times.

Table 1. Treatments for examining effects of density on growth of *V. humeralis* and *P. antipodarum* in field growth experiment

	Density treatments		
	400/m²	**800/m²**	**1200/m²**
Intraspecific effects			
Valvata humeralis	V	VV	-
Potamopyrgus antipodarum	P	PP	-
Interspecific effects			
V. humeralis + P. antipodarum		V+P	V+PP
P. antipodarum + V. humeralis		P+V	P+VV

Letters indicate treatment levels where V = *V. humeralis* at 400/m^2 and P = *P. antipodarum* at 400/m^2. Controls are represented by 'V' and 'P' treatments alone.

We constructed cages of clear modified plastic food storage containers (12 x 12 x 12 cm) with flow-through 0.5 mm Nitex vents on the top and two opposite sides to prevent snail movement in or out, but to allow water circulation. We covered the cage floor with local substrate (3-5 flat stones with a combined top surface area of approximately 130 cm^2) that was sifted and picked free of macroinvertebrates. We weighted cages to rest on the top surface of substrate and cage position was randomly assigned within a 5 X 5 m grid in the littoral zone at a depth of 0.5- 0.7 m. We assumed that cage effects would be minimal with this design because of the small size of gastropods used relative to the cages (1-3 mm versus 12 cm), the use of local substrate, and cage placement in local habitat.

We collected *V. humeralis* and *P. antipodarum* from the Vista reach using a Venturi suction dredge. We used juveniles of both species (*V. humeralis* < 2 mm and *P. antipodarum* < 3 mm) to capture growth potential and avoid measuring lack of growth associated with reproduction. We measured individual shell lengths to the nearest 0.01 mm with a digital micrometer on day 0 and then snails were randomly assigned to treatments. We did not mark individuals due to their small size, thus a randomly selected group of six individuals of each species in each treatment were used to calculate biomass. The experiment took place over the course of 10 days from August 11- 20, 2008. August is the month that the new annual cohort of *V. humeralis* appears and between August and September they grow rapidly (R. Newman, personal communication). We measured shell length as described above again on day

10 and converted shell lengths to ash free dry mass (AFDM) with an established length to mass equation for *P. antipodarum* (Hall et al. 2006). We developed our own length to mass equation for *V. humeralis* (Benke et al. 1999). The regression for *V. humeralis* was mass = $0.0868L^{3.1427}$ where mass is measured in mg (AFDM) and length (L) is measured in mm (n = 52, r^2 = 0.72) (Appendix A). Our *V. humeralis* mass includes shell mass while the *P. antipodarum* mass does not, thus biomass was relativized with initial biomass and relative growth changes were not compared between species. We calculated biomass specific growth rates as the ratio of after biomass to before biomass in ((mg/mg)/individual)/day modified from Cross and Benke 2002 as:

$$g = \frac{((M_t - M_0)/M0)/n}{t}$$

Where M_t is total gastropod biomass at the end of the experiment (mg AFDM), M_0 is the total initial gastropod biomass (mg AFDM), n is the number of individuals in the treatment used for the mean calculation (n = 6), and *t* is the duration of the experiment (days).

In the field experiment, we compared relative growth and mortality among treatments with a one-way multiple analysis of variance (MANOVA). If the MANOVA was significant, we used individual one-way analyses of variance (ANOVAs) for growth and mortality to explore independent effects of treatment type. If the effects in the ANOVAs were significant, we tested for significant differences among treatments with independent contrasts based on our experimental design. We compared the low-density intraspecific treatment to the medium density intraspecific, medium density interspecific, and high density interspecific treatments (Underwood 1997). We arcsine square root transformed percent mortality to meet assumptions of normality. Biomass values and variance met assumptions of normality and were not transformed.

We used changes in snail tissue biomass to calculate net effect sizes. We calculated effect size as the natural log of the ratio of treatment to control biomass change (Gurevitch and Hedges 2001; Kawai and Tokeshi 2007) as:

$$Net\ effect = ln(M_t/M_0)$$

Where M_t is total gastropod biomass at the end of the experiment (mg AFDM) and M_0 is the total initial gastropod biomass (mg AFDM). The control for all effect size calculations of a species was the low-density intraspecific treatment. To create three replicate net effect values for each comparison, we randomly selected three control and three treatment mean ratios (M_t/M_0) from the replicates and calculated a standard error. We compared the effects of treatment type on net effect with a one-way ANOVA. If the effect in the ANOVA was significant, we tested for significant differences among treatments with individual contrasts within species. We used an alpha value of 0.05 to confer significance and did not use dead snails in any statistical analysis.

Field Surveys

We examined biomass of *V. humeralis* from a population coexisting with *P. antipodarum* in the Vista reach to determine if growth results from our experiment were concordant with those of existing populations. We divided samples taken during the field surveys into categories matching those from the density and species proportions of the field experiment when possible. The semelparous life history of *V. humeralis* allowed us to use mean individual biomass from a sampling event as the mean individual biomass of the population on that sampling date. We selected one long-standing *V. humeralis* population that had been present in the Vista reach for at least six years from prior surveys (1998-2006) and had contained *P. antipodarum* this entire time.

The *V. humeralis* population was sampled monthly during the second week of each month from May through November of 2008 and May through August of 2009. Sampling months reflected the active period of the year for gastropods in the region when they can be found most readily (Cleland 1954; Hershey 1990; Lysne 2003; Lysne and Koetsier 2006b). We established a fixed transect in April which ran perpendicular to shore (perpendicular transect) through the greatest density of *V. humeralis* in the population. We marked the fixed transect on shore with rebar and revisited each month thereafter as the reference starting point for sampling. We began sampling each month by sampling along the perpendicular transect every 2 m moving out from the rebar and ceased when *V. humeralis* density decreased by 75% from the highest sample density. We placed another transect parallel to shore (parallel transect) through the greatest density of *V. humeralis* found while sampling along the perpendicular transect. We collected samples outward every 5 m along the parallel transect beginning at the intersection with the perpendicular transect. We ceased sampling along the parallel transect in both directions as described above.

We collected samples with a 0.25 m^2 plot, Venturi suction dredge, and SCUBA certified diver. We transported samples to the surface in flexible piping and passed through a 1 mm sieve bucket to remove sand and silt. We separated all *V. humeralis* and measured them from the anterior end of the shell across to the apex to the nearest 0.01 mm with digital calipers. We converted *V. humeralis* shell length to biomass (mg AFDM) as described previously. When samples were too large to search entirely, we thoroughly mixed contents and sub-sampled 50%. We considered samples devoid of *V. humeralis* and *P. antipodarum* after 30 min. of searching and we returned all sample contents to the plot where they were collected after measurements were taken. We counted abundance of other gastropod species present during 2009 sampling.

We compared mean individual *V. humeralis* biomass (size) from sample plots among categories matching the growth experiment with a one-way ANOVA in both years separately. If the effect in the ANOVAs was significant, we tested for significant differences among treatments with individual contrasts matching those of the field growth experiment (low density intraspecific compared to all other treatments). We used an alpha value of 0.1 to confer significance with the field survey data to balance error types and avoid being overly restrictive. We log transformed biomass to meet assumptions of normality in 2008 and 2009.

All measures of variability listed in parentheses are one standard error. We made no corrections for multiple independent contrasts following significant ANOVA effects because all contrasts were planned in the design of the experiments (Keppel and Wickens 2004). We performed all statistical procedures in SAS (9.3).

Results

Field Experiment

Mortality and relative growth varied significantly among treatments (Wilks' Lambda = 0.08, $F_{18,34} = 4.62$, $p < 0.0001$). Relative growth ($F_{9,18} = 3.14$, $p = 0.02$) and mortality ($F_{9,18} = 8.21$, $p < 0.0001$) were affected by treatment. *Valvata humeralis* growth was significantly greater in treatments containing *V. humeralis* and *P. antipodarum* at 400/m^2 each (total gastropod density 800/m^2) when compared to growth of *V. humeralis* alone at 400/m^2 ($F_{1,18}$ =

7.52, p = 0.01). The increased growth of *V. humeralis* with *P. antipodarum* disappeared at higher gastropod densities (Figure 1). Mortality of *V. humeralis* was greatest in the low-density intraspecific treatment and was marginally significantly different from mortality in the medium density intraspecific treatment ($F_{1,18} = 3.73$, p = 0.07) (Figure 1).

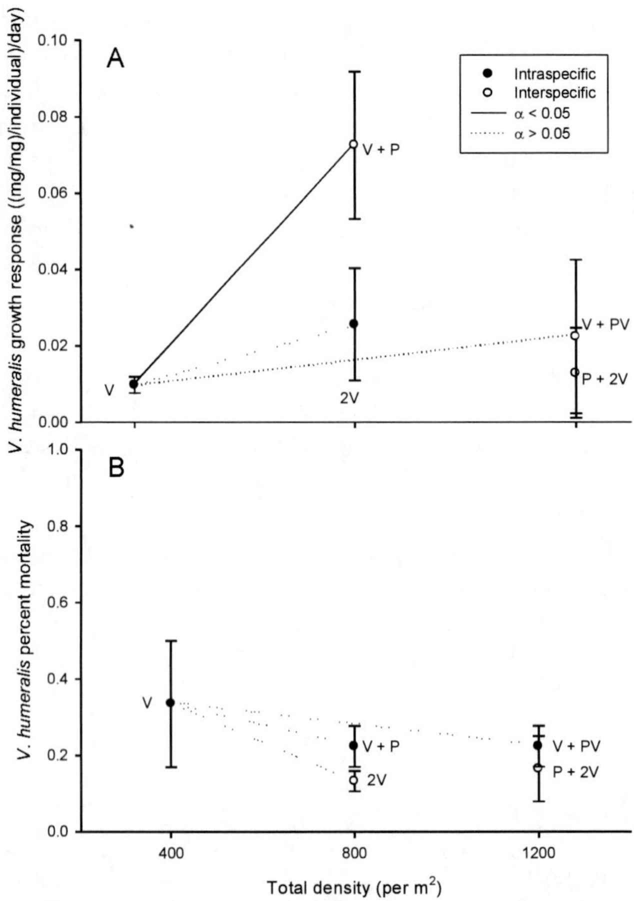

Figure 1. *Valvata humeralis* individual growth response (panel A) and mortality (panel B) in the five field experiment treatments. Filled circles represent intraspecific treatments and open circles represent interspecific treatments. The x-axis indicates total gastropod densities of low (400/m2), medium (800/m2), and high (1200/m2). The y-axis indicates mean growth response in ((mg/mg)/individual)/day. Lines represent independent contrasts with significance indicated in the legend. Error bars represent one stand error.

Although not statistically significant, *P. antipodarum* growth decreased as gastropod density increased in interspecific treatments except for the high-density V+2P treatment where growth matched the low density (Figure 2). Mortality of *P. antipodarum* was generally low and significantly higher in the medium density combined species treatment than the low-density intraspecific treatment ($F_{1,18} = 6.76$, 0.02) (Figure 2).

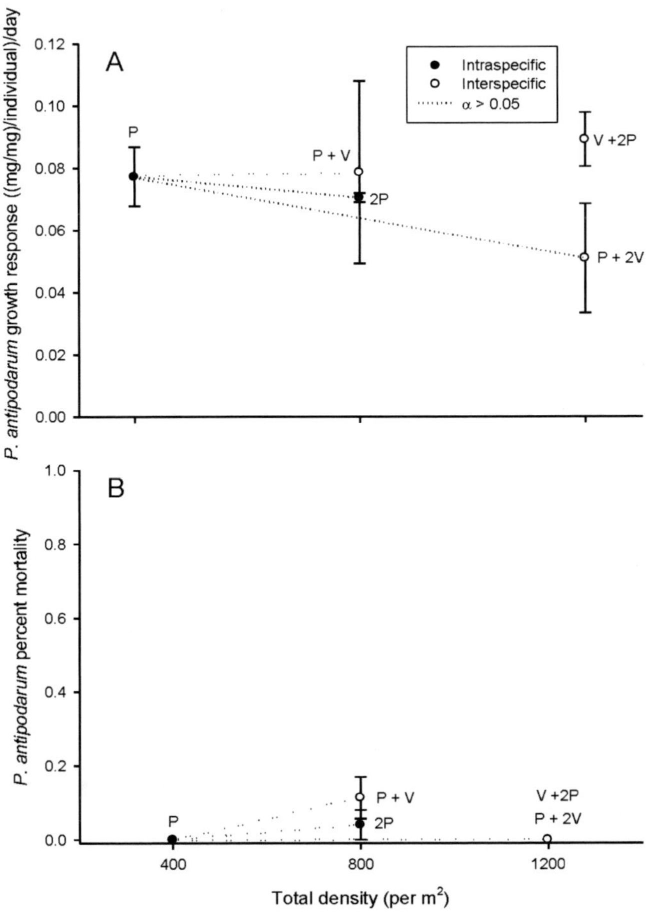

Figure 2. *Potamopyrgus antipodarum* individual growth response (panel A) and mortality (panel B) in the five field experiment treatments. Filled circles represent intraspecific treatments and open circles represent interspecific treatments. The x-axis indicates total gastropod densities of low (400/m^2), medium (800/m^2), and high (1200/m^2). The y-axis indicates mean growth response in ((mg/mg)/individual)/day. Lines represent independent contrasts with significance indicated in the legend. Error bars represent one stand error.

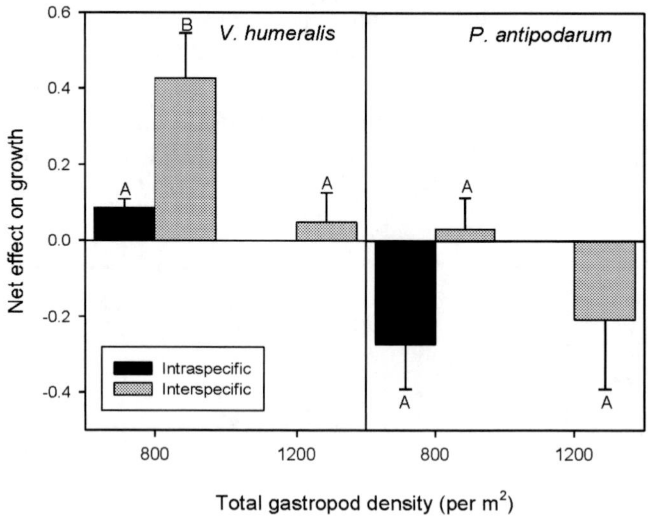

Figure 3. Mean intraspecific and interspecific net effects for *V. humeralis* and *P. antipodarum* from field growth experiment. Net effects represent the magnitude and direction of the effect of the interacting species on the target species as compared to the low density (400/m^2) intraspecific treatment for the target species. Those bars within species that share the same letters do not differ significantly with significance indicated in the legend. Error bars indicate one standard error.

Net effects varied significantly among intraspecific and interspecific treatments for both *V. humeralis* and *P. antipodarum* ($F_{5,12} = 4.98$, $p = 0.01$) (Figure 3). The interspecific net effect of *P. antipodarum* on *V. humeralis* was positive and significantly larger in the medium density than the high-density treatment ($F_{1,12} = 5.77$, $p = 0.03$). The net effect of *V. humeralis* on itself at 800/m^2 was slightly positive and significantly smaller than the positive interspecific net effect of *P. antipodarum* on *V. humeralis* at the same density ($F_{1,12} = 4.66$, $p = 0.05$). The net effect of *P. antipodarum* on itself was negative at 800/m^2 and not significantly different from the effects of *V. humeralis* on *P. antipodarum* at 800/m^2 ($F_{1,12} = 3.74$, $p = 0.08$) or 1200/m^2 ($F_{1,12} = 0.18$, $p = 0.68$).

Field Surveys

We collected 55 samples in 2008 and 46 samples in 2009 from the *V. humeralis* population. August was the only month during both years of sampling in which the densities and proportions of *V. humeralis* and *P. antipodarum* matched those from our growth experiment. Eleven samples from 2008 and eight samples from 2009 were used in the comparisons with the field experiment.

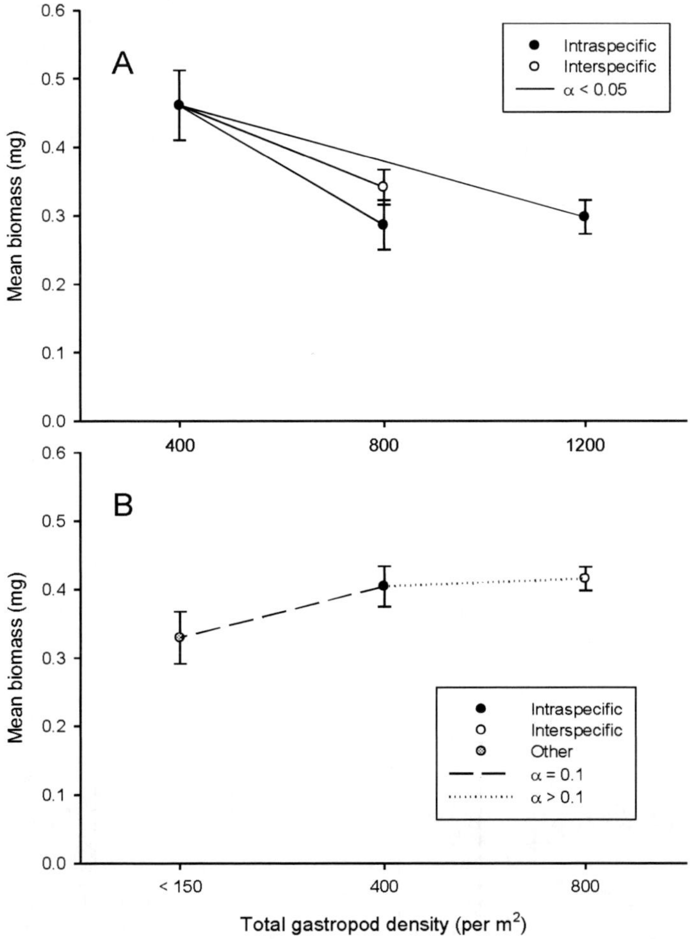

Figure 4. Mean biomass of *V. humeralis* from field samples collected in August of 2008 (panel A) and 2009 (panel B) by treatment categories from the field experiment. Filled circles represent intraspecific treatments, white circles represent interspecific treatments, and grey circles represent samples that did not fit into any of our experimental categories. The x-axis indicates target gastropod densities of low ($400/m^2$), medium ($800/m^2$), and high ($1200/m^2$). Lines represent independent contrasts with significance indicated in the legend. Error bars indicate one standard error.

Table 2. Sample splitting from August of 2008 and 2009 field surveys

No.	**Category**	**Field exp. density**	**Target density**	**Target proportion**	***V. humeralis*/m²**	***P. antipodarum*/m²**	**Total/m²**	**% *V. humeralis***	**n**
2008									
5	V	400/m²	< 500/m²	10:1 ± 1	276	32	308	0.90	4
9	V				438	54	492	0.89	6
7	VV	800/m²	> 500 and < 1000/m²	10:1 ± 1	848	0	848	1.00	16
10	VV				656	0	656	1.00	12
1	V+P			50:50 ± 10	328	172	500	0.66	82
11	V+P				320	152	472	0.68	5
2	VVV	1200/m²	> 1000/m²	10:1 ± 1	1280	16	1296	0.99	20
3	VVV				992	32	1024	0.97	16
4	VVV				1120	48	1168	0.96	19
2009									
3	V	400/m²	< 500/m²	10:1 ± 1	227	26	253	0.90	47
4	V+P	800/m²	> 500 and < 1000/m²	50:50 ± 10	344	216	560	0.61	84
5	V+P				126	192	318	0.40	26
7	V+P				184	216	400	0.46	33
8	V+P				88	128	216	0.41	10
2	Low other	na	> 20 and < 150/m²	na	60	61	121	0.50	15
6	Low other				12	120	132	0.09	3
9	Low other				40	64	104	0.38	5

Categories were designed to match the density and proportion of species from treatments in the field growth experiment. No. indicates sample number. Target density and target proportion describe sample splitting criteria. Species densities list actual number of each species and total density of both species per m². Sample number (n) indicates number of shells measured in a sample and used in analysis of variance (ANOVA). Samples numbers not listed had < 20 *V. humeralis*/m².

We split samples from August 2008 into four categories and samples from August 2009 into three categories (Table 2). In 2008, we measured the shell lengths of 180 *V. humeralis*. Treatment category was a significant predictor of size ($F_{3,177}$ = 2.23; p = 0.09). Matching our growth experiment, mean *V. humeralis* size in the V+P treatment was greater than those in the VV and VVV treatments although independent contrasts were not made because they were not planned. Contrary to our growth experiment, mean size in samples matching the V (n = 10) treatment was significantly greater than mean size in samples matching the VV (n = 28) ($F_{1,177}$ = 6.24, p = 0.01), V+P (n = 87) ($F_{1,177}$ = 4.56, p = 0.03), and VVV (n = 55) ($F_{1,177}$ = 5.42, p = 0.02) treatments (Figure 4). In 2009, we measured the shell lengths of 223 *V. humeralis*. Treatment category was a significant predictor of *V. humeralis* size ($F_{2,221}$ = 2.31; p = 0.10). Mean size in samples matching the V treatment (n = 47) were not significantly different from mean size in samples matching the V+P treatment (n = 153) ($F_{1,221}$ = 0.13, p = 0.72), but were significantly greater than the mean size in the lower density other category (n = 23) ($F_{1,221}$ = 2.71, p = 0.10) suggesting a benefit from the added *P. antipodarum* similar to our growth experiment (Figure 4).

Other species found in *V. humeralis* population samples in 2009 included *G. parvus*, *Lymnaidae* sp., *P. gyrina, Stagnicola* sp., *V. utahensis*, and *V. effusa.* Species other than *V. humeralis* and *P. antipodarum* were found at densities of < 40/m^2 in fewer than 25% of the samples except for *G. parvus* which was found at densities as high as 568/m^2.

Discussion

Our results provide evidence that interactions among native and invasive species can change in intensity and direction along invasive (benefactor) density gradients facilitating the native (beneficiary). In our growth experiment, *V. humeralis* exhibited increased growth in interspecific medium density treatments with *P. antipodarum* compared to growth in low density intraspecific treatments. The facilitative effect of *P. antipodarum* on *V. humeralis* growth was not apparent in the high-density interspecific treatment suggesting that the facilitation was directly related to the medium density of *P. antipodarum*. The net effect calculations mirrored these results showing an increase in the net effect of *P. antipodarum* on *V. humeralis* size when compared with the low-density intraspecific treatment and high density interspecific treatment. In contrast to *V. humeralis*, *P. antipodarum* experienced decreased growth with increasing density in the interspecific treatments but not intraspecific treatments. Intraspecific net effects of *P. antipodarum* were slightly greater than interspecific effects at the medium density suggesting that intraspecific competition may be greater than interspecific competition with *V. humeralis*.

Field surveys of *V. humeralis* size supported our experimental results on some levels but not others. Similar to our field experiment results, size in the medium interspecific treatment (hereafter referred to as facilitation treatment) in 2008 was larger than the medium density intraspecific or high-density intraspecific treatment although not statistically significant. Surveys in 2009 also matched our growth experiment in that *V. humeralis* size in samples that matched the facilitation treatment was greater than size in the low-density other category suggesting a benefit to the presence of added *P. antipodarum*. Yet, size in samples matching the low-density intraspecific *V. humeralis* treatment were either greater (2008) or the same (2009) as size in samples matching our facilitation treatment. Sample sizes for the low-density intraspecific treatment were low in 2008 and may not provide a robust estimate. Field survey results from both years indicate that the interaction between *V. humeralis* and *P. antipodarum* is influenced by other factors in a field setting such as abiotic conditions and neighboring species.

Several factors may have contributed to the differences we saw in our field experiment and field survey data. The quantity of food was standardized in our experiment, but likely fluctuated in the field setting. Difference in the abundance and nutritional quality of food among plots could have contributed to the muted facilitative effect of *P. antipodarum* on *V. humeralis* in the field survey. In addition, other gastropod species were present in the field surveys creating more complex community interactions. Most notably, *G. parvus* densities were greater at times than our target species, *V. humeralis* and *P. antipodarum*. The influence of *G. parvus* on our target species growth is not known, but likely affected our results. Lastly, mortality in the low-density intraspecific *V. humeralis* treatment was higher than in other treatments and may have contributed to the decreased growth in this treatment although no deceased gastropods were used in growth calculations.

Our research supports the hypothesis that some of the variation in the impact of *P. antipodarum* on native species among geographic locations may be explained by interaction changes along invader density gradients. The ability of *P. antipodarum* to reach extremely high densities suggests they would compete with native macroinvertebrates for habitat and food; however, prior research with *P. antipodarum* has demonstrated mixed responses. *Potamopyrgus antipodarum* have been found to dominate native macroinvertebrate assemblages in Australia (Quinn et al. 1998); however, when *P. antipodarum* densities were manipulated over brief temporal time scales (6 days) in previously invaded Australian streams, higher densities of *P. antipodarum* were positively correlated with native macroinvertebrate diversity (Schreiber et al. 2002). In a nearly opposite outcome, *P. antipodarum* densities were found to negatively affect macroinvertebrate colonization rates in the Greater Yellowstone Ecosystem (GYE) (Kerans et al. 2005); yet later surveys revealed broad-scale reductions in density of previously high density populations (B. Kerans and R. Richards, unpublished). Mixed results from previous research exploring the impact of *P. antipodarum* on native macroinvertebrates could have resulted from shifting interactions that promote coexistence at some invader densities but not others.

Our results agree with previous experiments indicating that *P. antipodarum* can facilitate other gastropod species as well as be facilitated by them under different conditions potentially affecting invader impact. For example, *P. antipodarum* presence was found to increase the fecundity of another invading snail, *Physella acuta*, when both snails were kept at equal densities (Cope and Winterbourn 2004) suggestive of reproductive facilitation. In an opposite outcome, Riley and colleagues (2008) found that *P. antipodarum* growth was facilitated by the presence *Pyrgulopsis robusta* potentially accelerating *P. antipodarum* impact in some settings. Our experimental results showed that growth of the native *V. humeralis* was facilitated by *P. antipodarum* when densities of the invader were moderate. While our field survey results were slightly different, they still indicated that the addition of *P. antipodarum* in plots did not decrease growth of *V. humeralis* indicating a lack of or low levels of competition. Combined, our experiment and field survey results show invader impact changing non-linearly with invader density and illustrate a mechanism of varied invader impact among locations and native species.

The mechanism of *P. antipodarum*'s facilitative effect on *V. humeralis* growth and the reason why it was density specific is uncertain. Previous research with gastropods has provided evidence of facilitation among species at varying densities (Brown 1982; Hershey 1990; Cope and Winterbourn 2004; Riley et al. 2008) with coprophagy frequently proposed as a potential mechanism (Schreiber et al. 2002). There is ample evidence that invertebrate herbivores can increase the quality of food around them by grazing (Hunter and Russell-Hunter 1983; Ruess and McNaughton 1984; Hillebrand et al.

2000; Liess and Hillebrand 2006; Liess and Kahlert 2007). Freshwater benthic grazers can increase epilithon nutrient content via excretion creating higher quality in areas where they graze (Hillebrand and Kahlert 2001; Hillebrand et al. 2004; Evans-White and Lamberti 2005). Schreiber and colleagues (2002) hypothesized that nutrients excreted by *P. antipodarum* provided food for other macroinvertebrates thereby increasing habitat suitability and increasing native species diversity. The benefit of increased food quality from *P. antipodarum* could be outweighed by increased competition as density of *P. antipodarum* increases potentially explaining why the facilitative effect of *P. antipodarum* on *V. humeralis* growth in our experiment was not apparent at higher densities. In other words, competitive interactions at higher densities may have dampened the facilitation seen at medium densities making it undetectable.

The implications of interactions shifting along biotic gradients are potentially substantial for invasive molluskan species. Introduced competitors are suggested as a driving force of extinction among global native freshwater mollusks (Lydeard et al. 2004). In North America, the persistence of native molluskan fauna has been threatened by the introductions of the Zebra mussel *Dreissena polymorpha*, the Quagga mussel *Dreissena rostriformis bugensis*, the Faucet snail *Bithynia tentaculata* (Jokinen 1992; Harman 2000), the Island apple snail *Pomacea insularum*, in addition to *P. antipodarum* (Lysne and Koetsier 2008; Riley et al. 2008). Introduced competitors may impact native Molluskan fauna through habitat alteration (Tucker 1994), competition for food and space (Murray et al. 1998; Baker and Levinton 2003; Haynes et al. 2005; Kerans et al. 2005; Hall et al. 2006), and resource depletion (Holland 1993; Nichols and Hopkins 1993; Caraco et al. 1997; Hall et al. 2003). Yet many questions remain regarding behavioral interactions that could affect the outcome of these interactions. Our results support the hypothesis that biotic and abiotic gradients may create unexpected population level impacts between native and invasive species (Dick 2008) potentially favoring native species.

Benefactor and beneficiary interactions that shift along biotic density gradients may also have widespread and potentially substantial implications for community dynamics. It has been increasingly acknowledged that shifting interactions can affect community structure and promote species coexistence (Choler et al. 2001; Peacor and Werner 2004; Gross 2008). Modeling efforts suggest that resource competitors can coexist in stable multi-species assemblages when species interactions shift from predominantly competitive to facilitative along a stress gradient even when the net interaction remains negative (Gross 2008). The facilitative species, or benefactor, may even create new interaction webs that indirectly affect other species within the community (Hacker and Gaines 1997). Indeed, research with freshwater bivalves suggests that species interactions may shift at higher abundances altering biodiversity affects (Allen and Vaughn 2011). If interaction shifts can occur along biotic gradients of species composition, shifting interactions may be much more common and influential in determining community structure than previously thought.

Past studies of shifting interactions between benefactor and beneficiary species have focused on abiotic drivers, but biotic gradients may be as important in structuring communities. We found the net effects between an invasive and native species shifted positively with increasing density of the invasive favoring the native's growth. Our research highlights the importance of considering biotic density gradients as drivers of species interaction shifts. Further, our research shows that non-linearity of interaction coefficients can affect invasive and native species interactions influencing the ecological impacts of invasive species. Further research is needed to explore mechanisms of biotic density gradients.

Conflict of Interest Statement

The authors state no conflict of interest in this work.

Funding

Funding for this work was provided by the U.S. Bureau of Reclamation and Montana State University, Bozeman, USA.

References

Agrawal, A. A., D. D. Ackerly, F. Adler, A. E. Arnold, C. Caceres, D. F. Doak, E. Post, P. J. Hudson, J. Maron, K. A. Mooney, M. Power, D. Schemske, J. Stachowicz, S. Strauss, M. G. Turner, and E. Werner. 2007. Filling key gaps in population and community ecology. *Frontiers in Ecology and the Environment* 5:145-152.

Allen, D. C. and C. C. Vaughn. 2011. Density-dependent biodiversity effects on physical habitat modification of freshwater bivalves. Ecology 92:1013-1019.

Alonso, A. and P. Castro-díez. 2008. What explains the invading success of the aquatic mud snail *Potamopyrgus antipodarum* (Hydrobiidae, Mollusca)? *Hydrobiologia* 614:107-116.

Baker, S. M., and J. S. Levinton. 2003. Selective feeding by three native North American freshwater mussels implies food competition with zebra mussels. *Hydrobiologia* 505:97-105.

Benke, A. C., A. D. Huryn, L. A. Smock, and J. B. Wallace. 1999. Length-mass relationships for freshwater macroinvertebrates in North America with particular reference to the southeasteran United States. *Journal of the North American Benthological Society,* 18:308-343.

Bertness, M. D. 1989. Intraspecific competition and facilitation in a northern acorn barnacle population. *Ecology* 70:257-268.

Bertness, M. D., and R. Callaway. 1994. Positive interactions in communities. *Trends in Ecology and Evolution* 9:191-193.

Bertness, M. D., G. H. Leonard, J. M. Levine, P. R. Schmidt, and A. O. Ingraham. 1999. Testing the relative contribution of positive and negative interaction in rocky intertidal communities. *Ecology* 80:2711-2726.

Bowler, P. A. 1991. The rapid spread of the freshwater hydrobiid snail *Potamopyrgus antipodarum* (Gray) in the middle Snake River, Southern Idaho. *Proceedings of the Desert Fishes Council* 21:173-182.

Bowler, P. A., and T. J. Frest. 1992. The non-native snail fauna of the Middle Snake River, Southern Idaho. *Proceedings of the Desert Fishes Council* 23:28-44.

Bronstein, J. L. 1994. Conditional outcomes in mutualistic interactions. *Trends in Ecology and Evolution* 9:214-217.

Brooker, R. W., and T. V. Callaghan. 1998. The balance between positive and negative plant interactions and its relationship to environmental gradients: a model. *Oikos* 81:196-207.

Brooker, R. W., F. T. Maestre, R. M. Callaway, C. L. Lortie, L. A. Cavieres, G. Kunstler, P. Liancourt, K. Tielborger, J. M. J. Travis, F. Anthelme, C. Armas, L. Coll, E. Corcket, S. Delzon, E. Forey, Z. Kikvidze, J. Olofsson, F. Pugnaire, C. L. Quiroz, P. Saccone, K. Schiffers, M. Seifan, B. Touzand, and R. Michalet. 2008. Facilitation in plant communities: the past, the present, and the future. *Journal of Ecology* 96:18-34.

Brown, K. M. 1982. Resource overlap and competition in pond snails: an experimental analysis. *Ecology* 63:412-422.

Bruno, J. F. and M. Bertness. 2001. Habitat modification and facilitation in benthic marine communities. Pages 201-218 in M. Bertness, S. D. Gaines, and M. Hay editors. *Marine Community Ecology.* Sinauer Associates, Sunderland, Massachusetts.

Bruno, J. F., J. J. Stachowicz, and M. D. Bertness. 2003. Inclusion of facilitation into ecological theory. *Trends in Ecology and Evolution* 18:119-126.

Bruno, J. F., J. D. Fridley, K. D. Bromberg, and M. D. Bertness. 2005. Insights into biotic interactions from studies of species invasions. Pages 13-40 *in* D. F. Sax, J. J. Stachowicz, and S. D. Gaines editors. *Species Invasions.* Sinauer Associates, Sunderland, Massachusetts.

Bulleri, F., C. Cristaudo, T. Alestra, and L. Benedetti-Cecchi. 2011. Crossing gradients of consumer pressure and physical stress on shallow rocky reefs: a test of the stress-gradient hypothesis. *Journal of Ecology* 99:335-344.

Callaway, R. M. 1995. Positive interactions among plants. *Botanical Review* 61:306-349.

Callaway, R. M., and L. R. Walker. 1997. Competition and facilitation: a synthetic approach to interactions in plant communities. *Ecology* 78:1958-1965.

Callaway, R. M., R. W. Brooker, P. Choler, Z. Kikvidze, C. J. Lortie, R. Michalet, L. Paolini, F. I. Pugnaire, B. Newingham, E. T. Aschehoug, C. Armas, D. Kikodze, and B. J. Cook. 2002. Positive interactions among alpine plants increase with stress. *Nature* 417:844-848.

Caraco, N. F., J. J. Cole, P. A. Raymond, D. L. Strayer, M. L. Pace, S. E. G. Findlay, and D. T. Fischer. 1997. Zebra mussel invasion in a large, turbid river: phytoplankton response to increased grazing. *Ecology* 78:588-602.

Choler, P., R. Michalet, and R. M. Callaway. 2001. Facilitation and competition on gradients in alpine plant communities. *Ecology* 82:3295-3308.

Cleland, D. M. 1954. A study of the habits of *Valvata piscinalis* (Muller) and the structure and function of the alimentary canal and reproductive system. *Proceedings of the Malacological Society of London* 30:167-203.

Collier, K. J., and M. J. Winterbourn. 2000. *New Zealand Stream Invertebrates: Ecology and Implications for Management*. New Zealand Limnological Society, Christchurch, New Zealand.

Cope, N. J., and M. J. Winterbourn. 2004. Competitive interactions between two successful molluscan invaders of freshwaters: an experimental study. *Aquatic Ecology* 38:83-91.

Cross, W. F., and A. C. Benke. 2002. Intra- and interspecific competition among coexisting lotic snails. *Oikos* 96:251-264.

Davis, M. A. 2003. Biotic globalization: does competition from introduced species threaten biodiversity? *Bioscience* 53:481-489.

Dick, J. T. A. 2008. Role of behavior in biological invasions and species distributions; lessons from interactions between the invasive *Gammarus pulex* and the native *G-duebeni* (Crustacea: Amphipoda). *Contributions to Zoology* 77:91-98.

Dick, J. T. A., M. E. Alexander, J. M. Jeschke, A. Ricciardi, H. J. MacIsaac, T. B. Robinson, S. Kumschick, O. L. F. Weyl, A. M. Dunn, M. J. Hatcher, R. A. Paterson, K. D. Farnsworth and D. M. Richardson. Advancing impact prediction and hypothesis testing in invasion ecology using a comparative functional response approach. *Biological Invasions* 16:735-753.

Dorgelo, J. 1987. Density fluctuations in populations (1982-1986) and biological observations of *Potamopyrgus jenkinsi* in two trophically differing lakes. *Hydrobiological Bulletin* 21:95-110.

Dorgelo, J. 1988. Growth in a freshwater snail under laboratory conditions in relation to eutrophication. *Hydrobiologia* 157:125-128.

Dorgelo, J., H. Meester, and C. van Velzen. 1995. Effects of diet and heavy metals on growth rate and fertility in the deposit-feeding snail *Potamopyrgus jenkinsi* (Smith) (Gastropoda: Hydrobiidae). *Hydrobiologia* 316:199-219.

Evans-White, M. A., and G. A. Lamberti. 2005. Grazer species effects on epilithon nutrient composition. *Freshwater Biology* 50:1853-1863.

Gates, K. K. 2012. Coexistence between a native (*Valvata humeralis*) and a non-native (*Potamopyrgus antipodarum*) gastropod in the middle Snake River, Idaho: implications for invasive species impact. Dissertation, Montana State University, Bozeman. Scholarworks catalog key 1943852.

Geist, J. A., J. L. Mancuso, M. M. Morin, K. P. Bommarito, E. N. Bovee, D. Wendell, B. Burroughs, M. R. Luttenton, D. L. Strayer, S. D. Tiegs. 2021. The New Zealand mud snail (*Potamopyrgus antipodarum*): autecology and management of a global invader. *Biological Invasions* https://doi.org/10.1007/s10530-021-02681-7.

Graff, P., M. R. Aguiar, and E. J. Chaneton. 2007. Shifts in positive and negative plant interactions along a grazing intensity gradient. *Ecology* 88:188-199.

Graff, P. and M. R. Aguiar. 2011. Testing the role of biotic stress gradient hypothesis. Processes and patterns in arid rangelands. *Oikos* 120:1023-1030.

Gross, K. 2008. Positive interactions among competitors can produce species-rich communities. *Ecology Letters* 11:929-936.

Gurevitch, J., and L. V. Hedges. 2001. Meta-analysis: combining the results of independent experiments. Editors S. M. Scheiner and J. Gurevitch *in Design and Analysis of Ecological Experiments,* pp. 347-369, 2nd edn. Oxford University Press, Oxford, UK.

Hacker, S. D., and S. D. Gaines. 1997. Some implications of direct positive interactions for community species diversity. *Ecology* 78:1990-2003.

Hall, R. O. Jr., J. L. Tank, and M. F. Dybdahl. 2003. Exotic snails dominate nitrogen and carbon cycling in a highly productive stream. *Frontiers in Ecology and the Environment* 1:407-411.

Hall, R. O. Jr., M. F. Dybdahl, and M. C. VanderLoop. 2006. Extremely high secondary production of introduced snails in rivers. *Ecological Applications* 16:1121-1131.

Hernandez, M. J. 1998. Dynamics of transitions between population interactions: a non-linear interaction alpha-function defined. *Proceedings: Biological Sciences* 265: 1433-1440.

Hershey, A. E. 1990. Snail populations in arctic lakes: competition mediated by predation? *Oecologia* 82:26-32.

Hillebrand, H., B. Worm, and H. K. Lotze. 2000. Marine microbenthic community structure regulated by nitrogen loading and grazing pressure. *Marine Ecology- Progress Series* 204:27-38.

Hillebrand, H., and M. Kahlert. 2001. Effect of grazing and nutrient supply on periphyton biomass and nutrient stoichiometry in habitats of different productivity. *Limnology and Oceanography* 46:1881-1898.

Hillebrand, H., G. de Montpellier, and A. Liess. 2004. Effects of macrograzers and light on periphyton stoichiometry. *Oikos* 106:93-104.

Holland, R. E. 1993. Changes in planktonic diatoms and water transparency in Hatchery Bay, Bass Island Area, Western Lake Erie since the establishment of the zebra mussel. *Journal of Great Lakes Research* 19:617-624.

Harman, W. N. 2000. Diminishing species richness of mollusks in Oneida Lake, New York State, USA. *Nautilus* 114(3):120-126.

Haynes, J. M., N. A. Trisch, C. M. Mayer and R. S. Rhyne. 2005. Benthic macroinvertebrate communities in southwestern Lake Ontario following invasion of

Dreissena and *Echinogammarus*: 1983-2000. *Journal of the North American Benthological Society* 24(1):148-167.

Hunter, R. D., and W. D. Russell-Hunter. 1983. Bioenergetic and community changes in intertidal aufwuchs grazed by *Littorina littorea*. *Ecology* 64:761-769.

Irving, A. D. and M. D. Bertness. 2009. Trait-dependent modification of facilitation on cobble beaches. *Ecology* 90:3042-3050.

Jokinen, E. 1992. The Freshwater Snails (Mollusca: Gastropoda) of New York State. The University of the State of New York, The State Education Department, The New York State Museum, Albany, New York 12230. 112 pp.

Kawai, T., and M. Tokeshi. 2004. Variable modes of facilitation in the upper intertidal: goose barnacles and mussels. *Marine Ecology Progress Series* 272:203-213.

Kawai, T., and M. Tokeshi. 2006. Asymmetric coexistence: bidirectional abiotic and biotic effects between goose barnacles and mussels. *Journal of Animal Ecology* 75:928-941.

Kawai, T., and M. Tokeshi. 2007. Testing the facilitation-competition paradigm under the stress-gradient hypothesis: decoupling multiple stress factors. *Proceedings of the Royal Society* B 274:2503-2508.

Keppel, G., and T. D. Wickens. 2004. *Design and analysis: a researcher's handbook.* Fourth Edition. Prentice Hall.

Kerans, B. L., M. F. Dybdahl, M. M. Gangloff, and J. E. Jannot. 2005. *Potamopyrgus antipodarum*: distribution, density, and effects on native macroinvertebrate assemblages in the Greater Yellowstone Ecosystem. *Journal of the North American Benthological Society* 24:123-138.

Kolar, C. S., and D. M. Lodge. 2001. Progress in invasion biology: predicting invaders. *Trends in Ecology and Evolution* 16:199-204.

Langenstein, S., and P. Bowler. 1991. On-going Macroinvertebrate analysis using the biotic condition index and the appearance of *Potamopyrgus antipodarum* (Gray) in Box Canyon Creek, Southern Idaho. *Proceedings of the Desert Fishes Council* 21:183-193.

Liess, A., and H. Hillebrand. 2006. Role of nutrient supply in grazer-periphyton interactions: reciprocal influences of periphyton and grazer nutrient stoichiometry. *Journal of the North American Benthological Society* 25:632-642.

Liess, A., and M. Kahlert. 2007. Gastropod grazers and nutrients, but not light, interact in determining periphytic algal diversity. *Oecologia* 152:101-111.

Loo, S. E., R. M. Nally, and P. S. Lake. 2007. Forecasting New Zealand mudsnail invasion range: model comparisons using native and invaded ranges. *Ecological Applications* 17:181-189.

Lydeard, C., R. H. Cowie, W. F. Ponder, A. E. Bogan, P. Bouchet, S. A. Clark, K. S. Cummings, T. J. Frest, O.Gardominy, D. G. Herbert, R. Hershler, K. E. Perez, B. Roth, M. Seddon, E. E. Strong, and F. G. Thompson. 2004. The global decline of nonmarine mollusks. *Bioscience* 54:321-330.

Lysne, S. J. 2003. The life history and ecology of two endangered Snake River gastropods: Utah valvata (*Valvata utahensis*; Call) and Idaho springsnail (*Pyrgulopsis idahoensis*; Pilsbry). M.S. Thesis, Boise State University, Idaho.

Lysne, S., and P. Koetsier. 2006b. The life history of the Utah (desert) Valvata, *Valvata utahensis*, in the Snake River, Idaho. *Journal of Freshwater Ecology* 21:285-291.

Lysne, S., and P. Koetsier. 2008. Comparison of desert valvata snail growth at three densities of the invasive New Zealand mud snail. *Western North American Naturalist* 68:103-106.

Maestre, F. T., I. Martinez, C. Escolar, and A. Escudero. 2009. On the relationship between abiotic stress and co-occurrence patterns: an assessment at the community level using soil lichen communities and multiple stress gradients. *Oikos* 118:1015-1022.

Murray, J., E. Murray, M. S. Johnson, and B. Clarke. 1988. The extinction of *Partula* on Moorea. *Pacific Science* 42:150-153.

Nichols, K. H., and G. J. Hopkins. 1993. Recent changes in Lake Erie (north shore) phytoplankton: cumulative impacts of phosphorus loading reductions and the zebra mussel introduction. *Journal of Great Lakes Research* 19:637-647.

Peacor, S. D., and E. E. Werner. 2004. How dependent are species-pair interaction strengths on other species in the food web? *Ecology* 85:2754-2763.

Quinn, G. P., P. S. Lake, and E. S. G. Schreiber. 1998. Colonization by lake benthos of hard substrata in the water column versus on the bottom. *Marine and Freshwater Research* 49:157-161.

Richards, D. C., L. D. Cazier, and G. T. Lester. 2001. Spatial distribution of three snail species, including the invader *Potamopyrgus antipodarum* in a fresh-water spring. *Western North American Naturalist* 6:375-380.

Riley, L. A., M. F. Dybdahl, and R. O. Hall. 2008. Invasive species impact: asymmetric interactions between invasive and endemic freshwater snails. *Journal of the North American Benthological Society* 27:509-520.

Ruess, R. W., and S. J. McNaughton. 1984. Urea as a promotive coupler of plant-herbivore interactions. *Oecologia* 63:331-337.

Schreiber, E. S. G., P. S. Lake, and G. P. Quinn. 2002. Facilitation of native stream fauna by an invading species? Experimental investigations of the interaction of the snail, *Potamopyrgus antipodarum* (Hydrobiidae) with native benthic fauna. *Biological Invasions* 4:317-325.

Schreiber, E. S. G., G. P. Quinn, and P. S. Lake. 2003. Distribution of an alien aquatic snail in relation to flow variability, human activities and water quality. *Freshwater Biology* 48:951-961.

Siegismund, H. R., and J. Hylleberg. 1987. Dispersal-mediated coexistence of mud snails (Hydrobiidae) in an estuary. *Marine Biology* 94:395-402.

Smit, C., C. Vandenberghe, J. den Ouden, and H. Muller-Scharer. 2007. Nurse plants, tree saplings and grazing pressure: changes in facilitation along a biotic environmental gradient. *Oecologia* 152:265-273.

Smit, C., M. Rietkerk, and M. J. Wassen. 2009. Inclusion of biotic stress (consumer pressure) alters predictions from the stress gradient hypothesis. *Journal of Ecology* 97:1215-1219.

Soliveres, S., P. Garcia-Palacios, A. P. Castillo-Monroy, F. T. Maestre, A. Escudero, and F. Valladares. 2011. Temporal dynamics of herbivory and water availability interactively modulate the outcome of a grass-shrub interaction in a semi-arid ecosystem. *Oikos* 120:710-719.

Straube, D., E. A. Johnson, D. Parkinson, S. Scheu, and N. Eisenhauer. 2009. Nonlinearity of effects of invasive ecosystem engineers on abiotic soil properties and soil biota. *Oikos* 118:885-896.

Tielborger, K., and R. Kadmon. 2000. Temporal environmental variation tips the balance between facilitation and interference in desert plants. *Ecology* 81:1544-1553.

Tucker, J. K. 1994. Colonization of unionid bivalves by the zebra mussel, *Dreissena polymorpha*, in pool-26 of the Mississippi River. *Journal of Freshwater Ecology* 9:129-134.

Underwood, A. J. 1997. *Experiments in Ecology: Their Logical Design and Interpretation Using Analysis of Variance*. Cambridge University Press, Cambridge, UK.

Villarreal-Barajas, T. and C. Martorell. 2009. Species-specific disturbance tolerance, competition and positive interactions along an anthropogenic disturbance gradient. *Journal of Vegetation Science* 20:1027-1040.

Went, F. W. 1942. The dependence of certain annual plants on shrubs in southern California deserts. *Bulletin of the Torrey Botanical Club* 69:100-114.

Williamson, M. H. 1996. *Biological Invasions*. Chapman & Hall, Padstow, Cornwall.

Winterbourn, M. J. 1970. The New Zealand species of *Potamopyrgus* (Gastropoda: Hydrobiidae). *Malacologia* 10:283-321.

Zaranko, D. T., D. G. Farara, and F. G. Thompson. 1997. Another exotic mollusk in the Laurentian Great Lakes: the New Zealand native *Potamopyrgus antipodarum* (Gray 1843) (Gastropod, Hydrobiidae). *Canadian Journal of Fisheries and Aquatic Sciences* 54:809-814.

Chapter 4

Competitive and Facilitative Mechanisms of a Biotic Interaction Gradient between Native and Invasive Aquatic Gastropods

Kiza K. Gates[1,2,*] **and Billie L. Kerans**[1,3]
[1]Montana State University, Bozeman, MT, USA
[2]Washington Department of Fish and Wildlife, WA, USA
[3]Retired

Abstract

In species that interact through both competition and facilitation, net interaction strength can change from negative to positive along abiotic or biotic gradients as the dominance of competition and facilitation change. We explored competition and facilitation between the aquatic gastropods *Valvata humeralis* (a native) and *Potamopyrgus antipodarum* (an invasive). To assess potential nutrient facilitation, we measured C:N and C:P ratios of *V. humeralis* and *P. antipodarum* body content, excretion, egestion, and field epilithon. To assess competition, we measured the effects of intraspecific and interspecific density and food availability on *V. humeralis* growth and mortality with a laboratory growth experiment. *Potamopyrgus antipodarum* juveniles had a lower ratio of body C:P than *P. antipodarum* adults and both ages of *V. humeralis* indicating that *P. antipodarum* juveniles have a higher phosphorus demand. We did not detect a difference between phosphorus excretion and egestion rates of *P. antipodarum* and *V. humeralis* indicating that facilitation by nutrient enhancement was unlikely. In the laboratory growth experiment, relative growth of *V. humeralis* during week one decreased with increasing density regardless of food quantity indicating that interference was the competitive mechanism. However, growth of *V. humeralis* was increased in high density interspecific treatments during the second week suggesting ontogeny influences the interaction between species. We

* Corresponding Author's Email: kizagates@gmail.com.

hypothesize that the facilitative effect of *P. antipodarum* on *V. humeralis* was caused by increased access to food and potentially access to food pre-conditioning by *P. antipodarum*. Our results provide insight into how interactions between native and invasive species can change with density and ontogeny.

Introduction

Organisms may interact in multiple ways causing the effects of one species on another to vary with changing conditions (Bronstein 1994; Bruno et al. 2003). The intensity and direction of the net interaction between species may change as the strength of individual interactions change along abiotic or biotic gradients. Net interaction strength can change direction along gradients of environmental stress (Bertness and Callaway 1994; Callaway and Walker 1997; Brooker and Callaghan 1998; Kawai and Tokeshi 2007; Brooker et al. 2008), biotic stress (Smit et al. 2007; Smit et al. 2009; Graff et al. 2007; Graff and Aguiar 2011), spatial scale (Bertness et al. 1999; Callaway et al. 2002), or through time (Tielborger and Kadmon 2000; Kawai and Tokeshi 2007). The mechanisms linked to each gradient differ but are all related to a tradeoff between different simultaneous interactions. In the case of species interactions that change along biotic stress gradients related strictly to the density of one species, the mechanisms are of particular interest because they may have profound implications for species coexistence.

Competition is undoubtedly the most studied interaction between species (Connell 1983; Schoener 1983; Gurevitch 1992). In fact, biologically similar species that overlap in resource use are often assumed to compete (MacArthur and Levins 1967). Classic competition theory suggests that biological interactions between competing species take either a direct form (interference) or an indirect form (exploitation) (Elton and Miller 1954). In interference, one species directly excludes the other species from access to a common resource (Elton and Miller 1954). In exploitation, one species indirectly prevents access to a common resource by consuming it and making it unavailable to other species (Elton and Miller 1954). While not mutually exclusive, interference and exploitation competition provide a framework for understanding how net interaction strength can change with the addition of a positive interaction.

There is considerable empirical evidence suggesting that facilitation may play a greater role in structuring net interactions than previously thought (Bertness and Callaway 1994; Stachowicz 2001; Bruno et al. 2003). As with competition, facilitation can have direct and indirect forms, which are frequently linked to enhancement of resources by one species. For example, habitat enhancement occurs when habitat quality for one species is increased directly by the presence of another. This form of facilitation has been commonly identified among sessile organisms such as plants (Went 1942) and intertidal marine organisms (Bertness 1989; Bertness et al. 1999; Kawai and Tokeshi 2004) where the presence of an additional species directly reduces exposure of neighboring species to environmental stress. In these examples, the positive effects of stress reduction in high stress environments outweigh the negative effects of resource competition and thereby shift the net interaction from negative to positive.

Facilitative interactions may also occur indirectly through food or nutrients causing net species interactions to change. For example, terrestrial herbivores may facilitate one another by creating access to better quality food through grazing or browsing (Vesey-Fitzgerald 1960; Bell 1970; McNaughton 1976). Herbivores may also enhance resources by increasing food quality through stoichiometric facilitation (Evans-White and Lamberti 2005; Hillebrand et al. 2000; Hillebrand and Kahlert 2001; Hillebrand et al. 2004; Hunter and Russell-Hunter 1983; Liess and Hillebrand 2006; Liess and Kahlert 2007; Ruess and McNaughton 1984). When an animal consumes a resource, it retains the nutrients needed for metabolic function and excretes or egests excess (Elser and Urabe 1999). Thus, differences in body stoichiometry (specifically carbon:nitrogen:phosphorus or C:N:P) among animals feeding on the same food source cause different rates of nutrient excretion and egestion (Elser and Urabe 1999; Evans-White and Lamberti 2005). Nutrient enrichment experiments simulating excretion and egestion indicate that the nutrient content of food resources can be increased through this process (Cross et al. 2003: Liess and Hillebrand 2006) leading to changing net interactions between organisms (Sommer 1992). Research with aquatic grazers has documented intraspecific facilitation where increased density of grazers decreased food quantity, but increased per capita food quality shifting the net interaction from negative to positive along an increasing density gradient (Sommer 1992). In this example, the indirect facilitation of increasing food quality outweighs competition for resources and thereby shifts the net interaction from negative to positive.

We explored the net interaction between the aquatic gastropods *Valvata humeralis* and *Potamopyrgus antipodarum* in a system where the former is native, and the latter is invasive. The net effect of *P. antipodarum* on *V. humeralis* shifts from negative to positive to negative along a gradient of increasing *P. antipodarum* density (Gates 2012). We designed two experiments to explore the facilitative and competitive mechanisms of their net interaction.

To explore the facilitative effect of *P. antipodarum* on *V. humeralis*, we investigated the potential of indirect habitat enhancement because we assumed that direct habitat enhancement was an unlikely facilitative mechanism between mobile gastropods. We measured the potential of *P. antipodarum* to enhance food quality for *V. humeralis* through nutrient recycling and or coprophagy (consumption egestion) by exploring the body content of essential nutrients for both species as well as excretion and egestion of essential nutrients. We also measured the nutrient content of food sources in the field to assess potential nutrient limitation. We hypothesized that body nutrient content would differ between species creating different nutrient requirements and that *P. antipodarum* would excrete and or egest more nutrients that were limiting to *V. humeralis* creating the potential for stoichiometric food facilitation.

In exploring competition, we assumed that interference competition would be more important than exploitation competition because aquatic gastropods have broad diets and are known to participate in coprophagy (López-Figueroa and Niell 1987; Brendelberger 1997). In this manner, food may be changed in quality after consumption but is still available as a food source and is thus never made unavailable to competitors (Lopez-Figueroa and Niell 1987; Brendelberger 1997). We therefore tested for interference competition by experimentally quantifying the net effects of *P. antipodarum* on *V. humeralis* along a gradient of gastropod density and species proportion while controlling food quantity on a per individual basis. Controlling food quantity per individual allowed us to infer a direct behavioral interaction if differences in growth occurred among treatments. We hypothesized that interference competition would cause *V. humeralis* growth to decrease linearly with increasing intraspecific density but would change non-linearly with increasing interspecific density based on previous experiments.

Methods

Stoichiometric Facilitation Experiment

We investigated stoichiometric food facilitation from the perspective of potentially limiting nutrients. We measured C:N and C:P ratios of *V. humeralis* and *P. antipodarum* body content to determine whether limiting nutrients differ between species. We determined juvenile and adult body nutrient content separately for both species because differences in nutrient excretion and egestion among animals may be more pronounced during different life stages (Elser et al. 1996).

We sampled adult *V. humeralis* and *P. antipodarum* from the Vista reach of the Middle Snake River, U.S.A. (Latitude 42°46'03", Longitude 112°52'46", RM 713) with a Venturi suction dredge and S.C.U.B.A. certified divers in July and August of 2009. The semelparous life cycle of *V. humeralis* allows a cohort to be followed through time and segregates the juvenile and adult life stages. Individuals sampled during any month are typically within 1 mm of the same size because they are from the same cohort (Gates 2012). Thus, we sampled adult and juvenile *V. humeralis* during the months when they were present in the population (adults in July and juveniles in August). We sampled *P. antipodarum* during July and August because they are iteroparous and both juveniles and adults were present both months. We collected fifty adult individuals of a species (*V. humeralis* > 2 mm and *P. antipodarum* > 3 mm), and fifty juveniles of a species (*V. humeralis* < 2 mm and *P. antipodarum* < 3 mm) and measured them to the nearest 0.01 mm with a digital micrometer. We stored snails on ice, returned them to the lab, wet-weighed them, de-shelled them, and froze them until further processing (Evans-White et al. 2005; Liess and Hillebrand 2005). We used pooled samples to obtain enough material for elemental analysis (seven samples per species and age group of 1.0- 1.5 mg/sample) (Evans-White et al. 2005). We freeze-dried pooled snail tissue, weighed it, and ground it to a fine powder with mortar and pestle. We analyzed subsamples for C and N in a CHN elemental analyzer and for total P (TDP) using the ascorbic acid method according to Rosemond (1993).

We measured the excretion and egestion rates of any nutrients that were significantly different between species in an additional experiment. Excretion and egestion were only measured on juveniles of both species because juveniles were used in the prior experiment (Gates 2012). We sampled *V.*

humeralis and *P. antipodarum* from the Vista reach with a Venturi suction dredge and S.C.U.B.A. certified divers in August of 2010. We collected fifty juvenile individuals of a species, measured them to the nearest 0.01 mm with a digital micrometer, and randomly pooled them into 5 replicates of 10 individuals per replicate. We placed pooled samples in 50 ml vials filled with river water filtered through glass-fiber filters to remove nutrient absorbing particles (Vanni et al. 2002). Vials contained a 1mm screen near the bottom to allow feces through and prevent coprophagy during the experiment (Whiles et al. 2009). We used five control samples to determine ambient N and P levels with filtered river water and no gastropods.

We synchronized the start of the egestion and excretion experiment among replicates. We held samples at river temperature for 2 hours in a dark cooler after which we filtered contents through pre-combusted glass-fiber filters to separate feces (egestion) and filtrate (excretion) (Vanni et al. 2002). We stored filters (containing fecal samples) and filtrate separately in vials, placed them on ice, and immediately returned them to the lab for processing (Vanni et al. 2002). We analyzed excretion samples as described above. We dried egestion samples (45 °C), weighed them, and analyzed them as described above.

We measured C:N and C:P of epilithon samples from the field to quantify nutrient availability and the potential for resource limitation. We randomly selected five stones from the *V. humeralis* population source in the Vista reach of the Snake River for epilithon analysis during August of 2010. We rinsed stones to remove invertebrates and scrubbed them with a nylon brush in pre-filtered river water until bare. We passed dislodged material through a pre-combusted glass fiber filter. We wrapped filters in tin foil, placed them on ice, and immediately returned them to the lab for processing. We analyzed filter subsamples for C and N in a CHN elemental analyzer and for total P (TP) as described above.

Total C, N, and P of *V. humeralis* and *P. antipodarum* tissue, epilithon, and laboratory Spirulina used as a food source are reported as percentages of dry masses and elemental ratios as molar ratios (C:N and C:P). We calculated mass specific excretion and egestion rates as the change in nutrient per unit time divided by the wet-mass of individuals (ug/mg wet-mass/hour) (Vanni et al. 2002). We obtained wet-mass by establishing a length to wet-mass regression for *V. humeralis* and *P. antipodarum* (Gates 2012).

Interference Competition Experiment

We assessed the effects of intraspecific and interspecific density and food availability on *V. humeralis* growth and mortality with a treatment/control replacement series experiment modified from Underwood (1997) and Cope and Winterbourn (2004). Treatments consisted of different species proportions at three densities (low (3 snails/cage or 400 snails/m^2), medium (6 snails/cage or 800 snails/m^2), and high (12 snails/cage or 1600 snails/m^2)) (hereafter referred to as density treatments) (Table 1). We selected densities to match field densities from prior surveys as well as prior field experiments (Gates 2012). We conducted the experiment over 21 days from September 4-25, 2010 in the Aquatic Sciences Laboratory at Montana State University, Bozeman, U.S.A. We designed the length of the experiment to be long enough to capture growth potential, but short enough to avoid reproduction by *P. antipodarum* (Tibbets et al. 2009).

Table 1. Laboratory experiment treatments for examining the effects of density, species composition, and food availability on growth and mortality of *V. humeralis*

	Low Food Density			**High Food Density**		
	400/m^2	**800/m^2**	**1600/m^2**	**400/m^2**	**800/m^2**	**1600/m^2**
Intraspecific effects						
V. humeralis	V	V + V	V + 3V	V	V + V	V + 3V
Interspecific effects						
V. humeralis + *P. antipodarum*		V + P	V + 3P		V + P	V + 3P

Letters indicate treatment levels where V = 3 *V. humeralis* individuals in the treatment container equating to 400/m^2 and P = 3 *P. antipodarum* individuals in the treatment container equating to 400/m^2.

We conducted the experiment in completely enclosed food storage containers (6 X 6 X 6 cm). We aerated water within treatments with individual air stones controlled by a MEDO® LA-45B air pump. We held containers in a living stream water bath of 20° C ± 2 to maintain constant temperature among all treatments. We monitored water temperature with HOBO Pro little v2® water temperature data loggers. We replaced water within treatment containers and cleaned containers after 7 and 14 days to prevent biofilm growth. We kept light within the laboratory on a 12:12 hr. light:dark regime.

We sampled *V. humeralis* two weeks prior to the start of the experiment from the Vista reach and held them in laboratory aquaria. We sampled juvenile

P. antipodarum from a laboratory-maintained Madison River, Montana, U.S.A. population because juveniles were not present in the Middle Snake River Vista reach at the time of *V. humeralis* sampling. We fed all gastropods ground *Spirulina platensis* (hereafter referred to as Spirulina) from The Vitamin Shoppe® while in holding tanks prior to the experiment (Krist and Lively 1998).

We measured individual shell lengths to the nearest 0.01 mm with a digital micrometer on day 0 and then randomly assigned individuals to treatments. We removed individuals from treatment containers on day 7, 14, and 21 to measure shell length with an ocular micrometer. We recorded any dead individuals and removed them from containers without replacement. We did not mark individuals; thus, we used a randomly selected group of three individuals of each species in each treatment to calculate biomass for day 0, 7, 14, and 21. We converted shell lengths to ash free dry mass (AFDM) with an established length to mass equation for *P. antipodarum* (Hall et al. 2006). We developed our own length to mass equation for *V. humeralis* (Benke et al. 1999). The regression for *V. humeralis* was mass = $0.0868L^{3.1427}$ where mass is measured in mg (AFDM) and length (L) is measured in mm ($n = 52$, $r^2 = 0.72$). Our *V. humeralis* mass includes shell mass unlike the established length to mass equation for *P. antipodarum*, thus we relativized biomass measurements by initial biomass and did not make comparisons between species. We calculated growth rates for individual weeks and for the entire experiment (comparing initial biomass to final biomass) in ((mg/mg)/individual)/day as:

$$g = \frac{((M_t - M_0)/M0)/n}{t}$$

Where M_t is total gastropod biomass at the end of the specified time period (mg AFDM), M_0 is the total initial gastropod biomass from the same specified time period (mg AFDM), n is the number of individuals in the treatment, and *t* is the duration of the specified time period within the experiment (days).

We controlled food quantity to provide two levels: 0.23 mg of food/snail in the low food quantity treatment and 0.46 mg food/snail in the high food quantity treatment (Krist et al. 2004; Tibbets et al. 2009). These quantities of food were used to test for differences in the intra- and interspecific interactions under unlimited and limited food defined in earlier work with *P. antipodarum* (Krist et al. 2004; Tibbets et al. 2009). We added food to treatment containers every other day at which time we turned off the air stone system for 4 hrs. to allow the food to settle. Combining the five density treatments levels (Table 1) and two food treatment levels created 10 total treatments. We adjusted food level in each treatment container every seven days to reflect mortalities because food was provided on a per individual basis. We replicated treatments four times. To make comparison between food used in the laboratory experiment and field food sources, we measured C:N and C:P of the Spirulina used in experiment. Spirulina C, N, and P were quantified as described above.

Statistical Analysis

We used a two-way (species and age) ANOVA to compare the C:N and C:P body content separately of *V. humeralis* and *P. antipodarum*. If the effects in the ANOVA were significant, we made all pair-wise comparisons with a Tukey correction. If there were significant differences in body nutrient content between species, we used two-sample t-tests to compare excretion and egestion rates of the nutrients between species. All measures of variability listed in parentheses are one standard error. Unless otherwise mentioned, we made no corrections for independent contrasts following significant ANOVA effects because all contrasts were planned in the design of the experiments (Keppel and Wickens 2004).

In the laboratory growth experiment, we used a two-way (density treatment and food level) multiple analysis of variance (MANOVA) to investigate differences in final relative biomass and mortality among treatments over the entire length of the experiment (hereafter referred to as total relative biomass and mortality). If the MANOVA was significant, we used individual two-way analyses of variance (ANOVAs) for growth and mortality with density and food to explore independent effects of treatment type. If the effects in the ANOVAs were significant, we tested for significant differences among treatments with independent contrasts based on our experimental design. Within food levels, the low density intraspecific treatment was compared to the medium density intraspecific, medium density interspecific, high density intraspecific, and high density interspecific treatments. We arcsine square root transformed percent mortality to meet assumptions of normality. Biomass values and variance met assumptions of normality.

To explore the effects of density treatment and food level on *V. humeralis* growth over time, we used a two-way (density treatment and food level) repeated-measures MANOVA (profile analysis). In our MANOVA, density treatment and food level were the main between subjects' factors and time was the main within subject factor (Von Ende 2001). Profile analysis analyzes the pattern of response of the within-subject factor and tests for three hypotheses of differences among treatments in parallelism (within subject interaction effects), flatness (within subject main effect), and levels (between subjects main effects) (Von Ende 2001). If the flatness and parallelism hypotheses are rejected, tests of the levels hypotheses are not reported (Von Ende 2001). The significance of the MANOVA was determined by the significance of the within subject effects of time (flatness hypothesis, indicated whether growth occurred over time), the time x density treatment interaction (parallelism hypothesis, indicated whether growth varied by density treatment over time), and the time x food level interaction (parallelism hypothesis, indicated whether growth varied by food level over time) all reported as Wilks' Lambdas (Von Ende 2001). If the MANOVA was significant, we used individual one-way ANOVAs for each seven-day interval (0-7, 8-14, 15-21) to explore independent effects of the density treatment and food level on growth with a Bonferroni corrected alpha ($\alpha = 0.02$) (Von Ende 2001). If the effects in the ANOVAs were significant, we tested for significant differences in growth among treatments with independent contrasts based on our experimental design. We made independent contrasts only within weeks. Biomass values and variance met assumptions of normality.

We used changes in snail tissue biomass to calculate net effect sizes. We calculated effect size as the natural log of the ratio of treatment to control biomass change (Gurevitch and Hedges 2001; Kawai and Tokeshi 2007) as:

$$Net\ effect = ln(M_t/M_0)$$

Where M_t is total gastropod biomass at the end of the experiment (mg AFDM) and M_0 is the total initial gastropod biomass (mg AFDM). The control for all *V. humeralis* effect size calculations was the low-density intraspecific treatment. We calculated a subset of three net effects for each treatment-control comparison by randomly selecting a control and treatment after/before biomass from the replicates and used this to calculate standard errors. We explored differences in the net effects among treatments with a one-way ANOVA. We did not use relative growth of *P. antipodarum* in any analyses, but it is reported. We did not use dead snails in statistical analyses. Unless otherwise mentioned, we used an alpha value of 0.05 to confer significance and performed all statistical procedures in SAS (9.3).

Results

Stoichiometric Facilitation Experiment

Valvata humeralis and *Potamopyrgus antipodarum* P body content (C:P) varied between species and ages ($F_{3,21} = 5.39$, $p = 0.007$) while N body content did not (C:N) ($F_{3,22} = 1.42$, $p = 0.26$) (Figure 1). Species and age alone were not significant predictors of body P content (species $F_{1,21} = 0.55$, $p = 0.47$, age $F_{1,21} = 3.24$, $p = 0.09$), but the species by age interaction was a significant predictor of body P content ($F_{1,21} = 12.1$, $p = 0.002$). *Potamopyrgus antipodarum* juveniles had a lower body C:P than *P. antipodarum* adults ($F_{1,21} = 14.68$, $p = 0.001$), *V. humeralis* juveniles ($F_{1,24} = 7.91$, $p = 0.01$), and marginally *V. humeralis* adults ($F_{1,24} = 3.4$, $p = 0.08$) (Figure 4). *Valvata humeralis* adults had a lower C:P than *P. antipodarum* adults ($F_{1,24} = 4.28$, $p = 0.05$) (Figure 1).

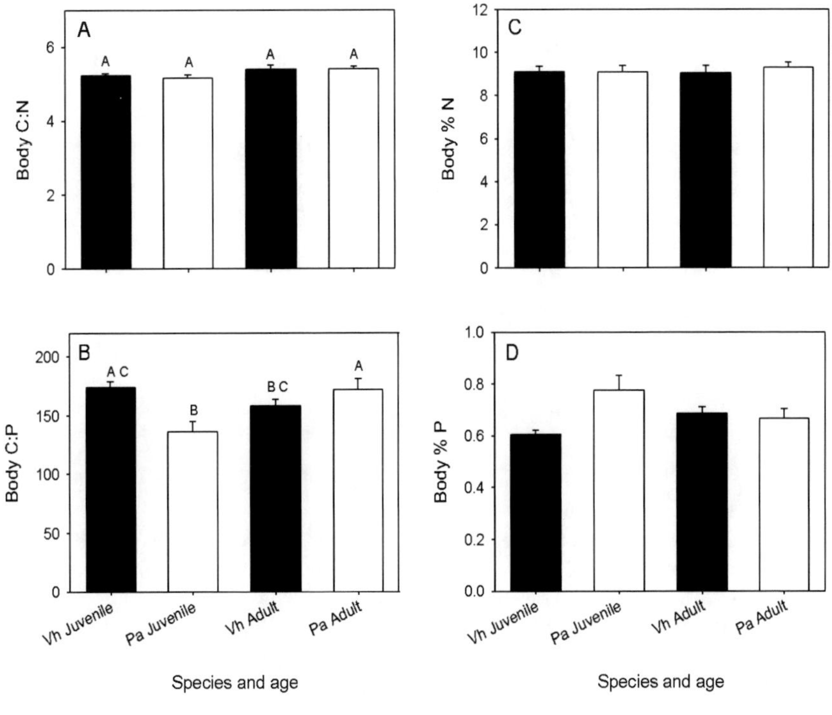

Figure 1. Ratios of C:N, C:P, percent N, and percent P for *V. humeralis* (black bars) and *P. antipodarum* (open bars) body tissue. Those bars with the same letter do not differ at the $\alpha = 0.05$ significance level. Error bars indicate one standard error.

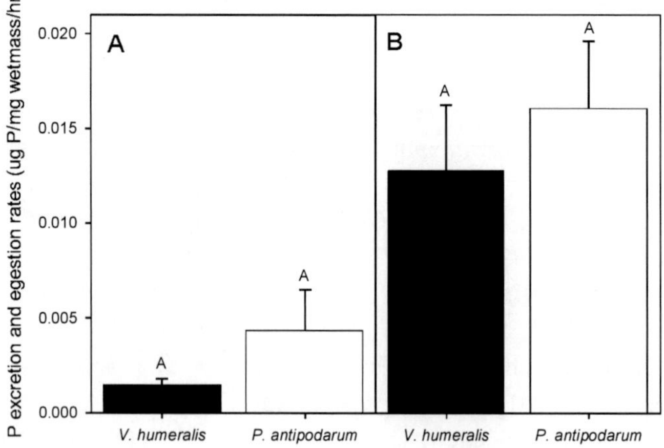

Figure 2. Phosphorus excretion (panel A) and egestion (panel B) rates for *V. humeralis* (black bars) and *P. antipodarum* (open bars). Those bars with the same letter do not differ at the $\alpha = 0.05$ significance level. Error bars indicate one standard error.

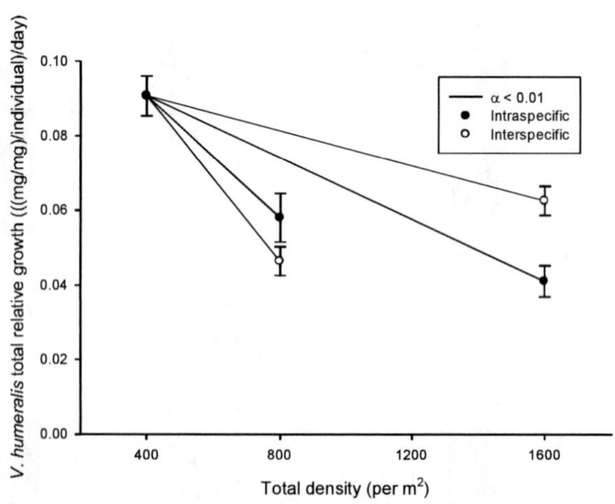

Figure 3. *Valvata humeralis* total relative growth over the entire 21 day experiment in the five density treatments. Food levels were combined because of lack of significance. Filled circles represent intraspecific treatments and open circles represent interspecific treatments. The x-axis indicates gastropod densities of low (400/m^2), medium (800/m^2), and high (1600/m^2). The y-axis indicates mean growth response in ((mg/mg)/individual)/day. Lines represent independent contrasts with significance values indicated in the legend. Error bars represent one stand error.

Phosphorus excretion and egestion rates did not differ between *P. antipodarum* and *V. humeralis* (excretion $t_8 = 1.31$, $p = 0.23$, egestion $t_8 = 0.66$,

$p = 0.52$) (Figure 2). Epilithon samples contained 17.81% C (0.90), 1.35% N (0.08), and 0.26% P (0.02) (C:N = 15.57 (0.40), C:P = 176.77 (10.23), C and N n = 10, P n = 10).

Interference Competition Experiment

Growth and mortality differed among density treatments (Wilks' Lambda = 0.30, $F_{8,58} = 5.89$, $p < 0.0001$), but we did not detect any differences in growth and mortality between food levels (Wilks' Lambda = 0.95, $F_{2,29} = 0.81$, $p = 0.46$) or the interaction between density treatment and food level (Wilks' Lambda = 0.74, $F_{8,58} = 1.2$, $p = 0.31$). Total relative growth was affected by the density treatment decreasing with increasing density ($F_{4,30} = 15.22$, $p < 0.0001$), but not by food level ($F_{1,30} = 1.58$, $p = 0.22$), or the interaction ($F_{4,30} = 0.65$, $p = 0.63$) (Figure 3). Mortality was not affected by density treatment ($F_{4,30} = 0.54$, $p = 0.70$), food level ($F_{1,30} = 0.20$, $p = 0.67$), or the interaction ($F_{4,30} = 1.77$, $p = 0.16$). Total relative growth of *V. humeralis* was significantly greater in the low density intraspecific treatment than all others (Figure 3). Total relative growth of *P. antipodarum* was 0.29 (0.08) ((mg/mg)/individual)/day in the low food medium density interspecific treatment, 0.28 (0.05) ((mg/mg)/individual)/day in the high food medium density interspecific treatment, 0.20 (0.03) ((mg/mg)/individual)/ day in the low food high density interspecific treatment, and 0.16 (0.11) ((mg/mg)/individual)/day in the high food high density interspecific treatment.

In the repeated-measures analysis, *V. humeralis* grew over time (Wilks' Lambda = 0.03, $F_{3,28} = 269.73$, $p < 0.0001$), density treatment affected growth over time (Wilks' Lambda = 0.34, $F_{12,74.37} = 3.10$, $p = 0.001$), and food level affected growth over time (Wilks' Lambda = 0.71, $F_{3,28} = 3.78$, $p = 0.02$). There was not a significant interaction of time, density treatment, and food level (Wilks' Lambda = 0.75, $F_{12,74.37} = 0.70$, $p = 0.74$). Within the three weeks of the experiment, the time effect was significant indicating that *V. humeralis* grew every week (Table 2). The density treatment was a significant predictor of growth in the first and third week, but not the second. Food level was only significant in week three and the interaction of density treatment and food level was not significant during any week. Relative growth of *V. humeralis* from days 1-7 was significantly larger in the low density intraspecific treatment than all others, but from days 8-14 was marginally significantly larger than all others in the high density interspecific treatment (Figure 4). Relative growth of *V. humeralis* from days 15-21 was generally greater in high food level treatments with the highest growth occurring in the low density, high food intraspecific treatment (Figure 4).

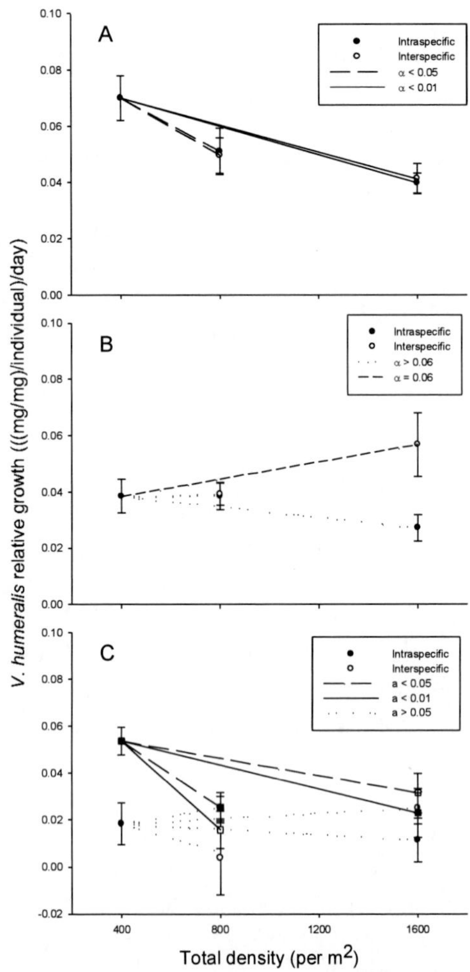

Figure 4. *Valvata humeralis* individual growth response in the five treatments during day 0-7 (panel A), 8-14 (panel B), and 15-21 (panel C). Filled circles represent intraspecific treatments and open circles represent interspecific treatments. The x-axis indicates gastropod densities of low (400/m^2), medium (800/m^2), and high (1600/m^2). The y-axis indicates mean growth response in ((mg/mg)/individual)/day. Food quantity was only significant in week 3 (panel C) where squares indicate high food and circles represent low food. Lines represent independent contrasts with significance values indicated in the legend. Independent contrasts were made within week among treatments types that were significant during that week. Error bars represent one stand error.

Table 2. Results from repeated measures two-way analysis of variance (ANOVA) with density and food level for days 0-7, 8-14, and 15-21 of the laboratory growth experiment

	Source	**DF**	**ANOVA SS**	**Means Square**	**F**	**p**
Days 0-7						
	Time	1	4.97	4.97	293.05	*< 0.0001*
	Density x Time	4	0.23	0.06	3.33	*0.02*
	Food x Time	1	0.02	0.02	1.06	0.31
	Density*Food*Time	4	0.05	0.01	0.78	0.54
	Error	30	0.51	0.02		
Days 8-14						
	Time	1	3.14	3.14	159.76	*< 0.0001*
	Density x Time	4	0.18	0.04	2.27	0.09
	Food x Time	1	0.02	0.02	0.79	0.38
	Density*Food*Time	4	0.02	0.01	0.2	0.94
	Error	30	0.59	0.02		
Days 15-21						
	Time	1	1.08	1.08	96.73	*< 0.0001*
	Density x Time	4	0.15	0.04	3.31	*0.02*
	Food x Time	1	0.08	0.08	6.89	*0.01*
	Density*Food*Time	4	0.07	0.02	1.59	0.2
	Error	30	0.34	0.01		

Significance was assessed with a Bonferroni corrected $\alpha = 0.02$ and significant p-values are italicized.

Net effects varied significantly among treatments with a trend of decreasing overall effects in high density interspecific treatments ($F_{7,16} = 7.72$, $p = 0.0004$) (Figure 5). Intraspecific and interspecific net effects were more negative at the lower food level within treatments. The lowest net effect occurred at the medium density (800/m^2) intraspecific high food treatment and the greatest net effect occurred in the high density (1600/m^2) low food intraspecific treatment.

Spirulina used in the laboratory experiment contained 42.54% C (1.80), 8.48% N (0.38) (C:N = 5.83 (0.02) (C and N n = 5). The quantity of P within the Spirulina used in our growth experiment was too high for our laboratory calibrations and was thus obtained from the manufacturer (4.0% P and C:P = 27.31).

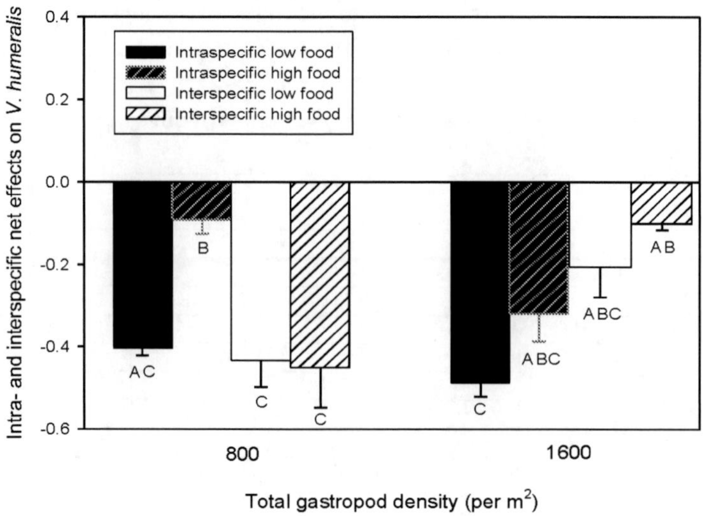

Figure 5. Mean intraspecific and interspecific net effects for *V. humeralis* from laboratory growth experiment. Net effects represent the magnitude and direction of the effect of the interacting species on the target species as compared to the low density (400/m^2) intraspecific treatment for the target species. Those bars with the same letter do not differ at the $\alpha = 0.05$ significance level. Error bars indicate one standard error.

Discussion

Our results suggest that the intensity of a net interaction between species can change in ways not predicted by community ecology theory due to changes in the multiple interactions that make up the net interaction. Total relative growth of *V. humeralis* decreased with increasing gastropod density regardless of the per individual food provided indicating that direct behavioral interactions were responsible for the decrease in growth rate with increasing density (interference competition). Yet, contrary to what would be predicted by theory, the intensity of competition between *V. humeralis* and *P. antipodarum* decreased with increasing *P. antipodarum* density. *Valvata humeralis* growth was increased in the high density interspecific treatments as compared to the high density intraspecific treatment. We hypothesize that *V. humeralis* competed with but was also facilitated by *P. antipodarum* at this density, changing the intensity of the net interaction.

Our results also provide insight into how net interactions between species can change with ontogeny creating non-linear interactions over time. Previous research with *V. humeralis* and *P. antipodarum* found a net interaction shift from competition, to facilitation, back to competition with increasing density of *P. antipodarum* (Gates 2012). While our current growth results do not directly match these, we saw evidence of a similar net interaction change through the course of our experiment. Growth potential was greatest from days 0-7 during which time growth rates decreased linearly with increasing total gastropod density regardless of species combinations and food levels. However, when growth potential decreased during days 8-14, there was a facilitative effect of *P. antipodarum* on *V. humeralis* growth at the highest density. By days 15-21, growth potential equalized among treatments as gastropods reached adult size and food quantity became a significant predictor of growth rate with higher growth in high food treatments. The importance of the facilitation that occurred during the second week of the experiment was confirmed in the total growth results. Growth of *V. humeralis* was greater in the high density interspecific treatment than the intraspecific treatment although an independent contrast was not made because it was not planned. The change in net effect seen through time in our experiment indicates that the facilitation of *V. humeralis* by *P. antipodarum* is related to *V. humeralis* ontogeny and may only become apparent once juveniles have begun decreasing in growth. The difference in shifting interactions seen between our laboratory experiment and previous field experiments may thus be related to the growth potential of *V. humeralis*. While juveniles were used in both experiments, it is likely that individuals varied slightly in age and growth potential among experiments.

The facilitation of *V. h*umeralis growth in the presence of *P. antipodarum* may have been caused by increased access to food, increased quality of food, or a combination thereof. The high quality of Spirulina food used in the growth experiment suggests that *V. humeralis* did not experience nutrient limited growth (N or P limitation), but it is possible that the nutrients within the Spirulina were packaged in a way that decreased accessibility. The Spirulina food source used in our experiment was a filamentous cyanobacterium. While it has been commonly used as a food source for gastropods in laboratory settings (Krist and Lively 1998; Tibbets et al. 2009), the precise digestion process is unknown. If nutrient packaging within the Spirulina limited access to nutrients, digestion by *P. antipodarum* could have made the nutrients more accessible, and thus higher in quality for *V. humeralis*. Coprophagy among gastropods is a viable strategy for further acquisition of nutrients within food that is not completely or efficiently processed by one digestion (Brendelberger 1997). It is also possible that nutrient levels within the lab Spirulina were too high (Boersma and Elser 2006) and that pre-digestion by *P. antipodarum* decreased the nutrient content to a more optimum level. In addition to potential changes in food quality caused by *P. antipodarum* digestion, *V. humeralis* may have also had greater access to the Spirulina food through interference behavior. Chemical cues released by *P. antipodarum* may have also affected growth of *V. humeralis* (Dillon 2000), although water within treatment cages was replaced every seven days reducing the potential impacts of chemical cues.

The density and species proportions at which growth of *V. humeralis* was facilitated by *P. antipodarum* differed somewhat from previous research perhaps due to variation in abiotic conditions. A previous field growth experiment found the facilitation of *V. humeralis* by *P. antipodarum* occurred at 800/m^2 total gastropod density and 1:1 *V. humeralis*:*P. antipodarum* proportion while our current results found the facilitation occurred at 1600/m^2 and 1:3 proportion (Gates 2012). It is important to note that cage sizes differed between experiments and while the density per area at which the facilitation occurred differed in the lab and field, the actual number of gastropods in the treatment where the facilitation occurred were similar between experiments (previous experiment 6 *V. humeralis* individuals and 6 *P. antipodarum* individuals, current experiment 3 *V. humeralis* individuals and up to 9 *P. antipodarum* individuals). This suggests that the interaction may be cued more by number of individual gastropods interacting and less by small differences in the size of the area that they interact within. The differences we saw in density and proportion creating the interaction shift may also be related to other factors intertwined with the density gradient. To examine the biotic density gradient in the current growth experiment, we controlled for abiotic gradients by conducting the experiment in a low abiotic stress environment. For example, the food provided in the lab experiment was much higher in limiting nutrients than the epilithon in the field and was continuously replenished reducing the potential for nutrient or food limited growth. The conditions in the laboratory were also optimal for growth with very small temperature swings from day to night and light levels lower than that of direct sunlight. A recent study investigating growth of *P. antipodarum* found that foraging was increased in low light treatments potentially contributing to increased growth (Liess and Lange 2011) suggesting that the low light conditions in our laboratory (compared to natural sunlight) may have increased growth rates in general. Decreasing environmental stress combined with increasing food quality and continually renewing food quantity in the current experiment could have pushed the facilitative effect to a higher density where competition was greater.

Our results agree with community ecology theories of species coexistence and resource overlap. Growth patterns from our experiment indicate that *V. humeralis* may coexist with *P. antipodarum* because interspecific competition with the invader is not as strong as intraspecific competition with itself. Previous research with *P. antipodarum* and native isopods in the Columbia River similarly found that interspecific competition with *Gnorimosphaeroma insulare* was weaker than intraspecific competition between native isopods suggesting a mechanism of coexistence (Brenneis et al. 2010). Additionally, our food level treatment became significant in the last week of the study when individual size had increased. Increasing food quantity decreased the intra- and interspecific competitive effects seen in the last week of our experiment indicating that competition was for space and food.

Results from our stoichiometric experiment suggest that *P. antipodarum* may egest and excrete slightly more phosphorus than *V. humeralis*, but it is likely not enough to be the sole facilitative mechanism in this or previous experiments. The Spirulina used in our experiment contained enough phosphorus that limitation for either species was highly unlikely (Tibbets et al. 2009) although the accessibility of these nutrients within the Spirulina is uncertain as is the potential excess of nutrients (Boersma and Elser 2006). The C:P of epilithon from field samples was also greater than that thought to limit *P. antipodarum* growth and fecundity (176 compared to ~270) indicating that phosphorus limitation at our field site is not likely (Tibbets et al. 2009). It is of interest that *P. antipodarum* juveniles require more phosphorus (higher body content) than *V. humeralis* juveniles suggesting that *P. antipodarum* may become phosphorus limited as juveniles before *V. humeralis.* As adults, the pattern was switched and *V. humeralis* adults had significantly higher demand for phosphorus than *P. antipodarum* adults as may be expected for a rapidly growing invasive species with high protein synthesis rates (Elser et al. 1996). These results suggest that *P. antipodarum* growth and population expansion may be limited by juvenile growth in some locations contributing to varying impact of this invasive species. These results are consistent with a previous meta-analysis suggesting that aquatic native species may out-compete invasive species under low nutrient condition where the invader's growth is limited (González et al. 2010).

Our results confirm that the net effects of one species on another can change in intensity along biotic density gradients because of changes in the interactions that sum to create the net interaction. In the current study, species interacted negatively through interference competition and positively through facilitated access to higher quantity and/or quality of food. The intensity of interspecific competition in our experiment decreased as the density of the competitor increased potentially because *V. humeralis* was able to physically defend areas where food settled and benefit indirectly through acquisition of food resources. While the interaction change occurred along a density gradient, environmental conditions including food availability and temperature affected the densities at which the shift occurred as compared to previous work (Gates 2012). These results are concordant with previous biotic gradient research indicating that biotic and abiotic interaction gradients can operate concurrently (Irving and Bertness 2009; Bulleri et al. 2011).

Our results have significant implication for the management of invasive species and the role of receiving native communities in the invasion process. Competition with the native community is already recognized as a potential influence during the invasion process that may prevent progression of the invasion ultimately influencing impact (Bertness and Callaway 1994; Stachowicz 2001; Price and Morin 2004). Our results suggest further that the impact of an invader on the native community may be a complex interplay between invader density, native species behavior, and environmental conditions. The weight of each of these components likely differs among invasions sites creating context specificity.

Conflict of Interest Statement

The authors state no conflict of interest in this work.

Funding

Funding for this work was provided by Montana State University and the Montana Water Center.

References

Bell, R. H. V. 1971. A grazing ecosystem in the Serengeti. *Scientific American* 224:86-93.

Benke, A. C., A. D. Huryn, L. A. Smock, and J. B. Wallace. 1999. Length-mass relationships for freshwater macroinvertebrates in North America with particular reference to the southeastern United States. *Journal of the North American Benthological Society* 18:308-343.

Bertness, M. D. 1989. Intraspecific competition and facilitation in a northern acorn barnacle population. *Ecology* 70:257-268.

Bertness, M. D., and R. Callaway. 1994. Positive interactions in communities. *Trends in Ecology and Evolution* 9:191-193.

Bertness, M. D., G. H. Leonard, J. M. Levine, P. R. Schmidt, and A. O. Ingraham. 1999. Testing the relative contribution of positive and negative interaction in rocky intertidal communities. *Ecology* 80:2711-2726.

Boersma, M., and J. J. Elser. 2006. Too much of a good thing: on stoichiometrically balanced diets and maximal growth. *Ecology* 87:1325-1330.

Brendelberger, H. 1997. Coprophagy: a supplementary food source for two freshwater gastropods? *Freshwater Biology* 38:145-157.

Brenneis, V. E. F., A. Sih, and C. E. De Rivera. 2010. Coexistence in the intertidal: interactions between the non-indigenous New Zealand mud snail *Potamopyrgus antipodarum* and the native estuarine isopod *Gnorimosphaeroma insulare*. Oikos 119:1755-1764.

Bronstein, J. L. 1994. Conditional outcomes in mutualistic interactions. *Trends in Ecology and Evolution* 9:214-217.

Brooker, R. W., and T. V. Callaghan. 1998. The balance between positive and negative plant interactions and its relationship to environmental gradients: a model. *Oikos* 81:196-207.

Brooker, R. W., F. T. Maestre, R. M. Callaway, C. L. Lortie, L. A. Cavieres, G. Kunstler, P. Liancourt, K. Tielborger, J. M. J. Travis, F. Anthelme, C. Armas, L. Coll, E. Corcket, S. Delzon, E. Forey, Z. Kikvidze, J. Olofsson, F. Pugnaire, C. L. Quiroz, P. Saccone, K. Schiffers, M. Seifan, B. Touzand, and R. Michalet. 2008. Facilitation in plant communities: the past, the present, and the future. *Journal of Ecology* 96:18-34.

Bruno, J. F., J. J. Stachowicz, and M. D. Bertness. 2003. Inclusion of facilitation into ecological theory. *Trends in Ecology and Evolution* 18:119-126.

Bulleri, F., C. Cristaudo, T. Alestra, and L. Benedetti-Cecchi. 2011. Crossing gradients of consumer pressure and physical stress on shallow rocky reefs: a test of the stress-gradient hypothesis. *Journal of Ecology* 99:335-344.

Callaway, R. M., and L. R. Walker. 1997. Competition and facilitation: a synthetic approach to interactions in plant communities. *Ecology* 78:1958-1965.

Callaway, R. M., R. W. Brooker, P. Choler, Z. Kikvidze, C. J. Lortie, R. Michalet, L. Paolini, F. I. Pugnaire, B. Newingham, E. T. Aschehoug, C. Armas, D. Kikodze, and B. J. Cook. 2002. Positive interactions among alpine plants increase with stress. *Nature* 417:844-848.

Connell, J. H. 1983. On the prevalence and relative importance of interspecific competition: evidence from field experiments. *The American Naturalist* 122:661-695.

Cope, N. J., and M. J. Winterbourn. 2004. Competitive interactions between two successful molluscan invaders of freshwaters: an experimental study. *Aquatic Ecology* 38:83-91.

Cross, W. F., J. P. Benstead, A. D. Rosemond, and J. B. Wallace. 2003. Consumer-resource stoichiometry in detritus-based streams. *Ecology Letters* 6:721-732.

Dillon, J. T. 2000. *The ecology of freshwater molluscs.* Cambridge University Press, New York.

Elser, J. J., D. R. Dobberfuhl, N. A. MacKay, and J. H. Schampel. 1996. Organism size, life history, and N:P stoichiometry. *Bioscience* 46:674-684.

Elser, J. J., and J. Urabe. 1999. Stoichiometry of consumer-driven nutrient recycling: theory, observations, and consequences. *Ecology* 80:735-751.

Elton, C., and R. S. Miller. 1954. The ecological survey of animal communities: with a practical system of classifying habitats by structural characters. *Journal of Ecology,* 42:460-496.

Evans-White, M. A., and G. A. Lamberti. 2005. Grazer species effects on epilithon nutrient composition. *Freshwater Biology* 50:1853-1863.

Evans-White, M. A., R. S. Stelzer, and G. A. Lamberti. 2005. Taxonomic and regional patterns in benthic macroinvertebrate elemental composition in streams. *Freshwater Biology* 50:1786-1799.

Gates, K. K. 2012. *Coexistence between a native (Valvata humeralis) and a non-native (Potamopyrgus antipodarum) gastropod in the middle Snake River, Idaho: implications for invasive species impact.* Dissertation, Montana State University, Bozeman. Scholarworks catalog key 1943852.

González, A. L., J. S. Kominoski, M. Danger, S. Ishida, N. Iwai, and A. Rubach. 2010. Can ecological stoichiometry help explain patterns of biological invasions? *Oikos* 119:779-790.

Graff, P., M. R. Aguiar, and E. J. Chaneton. 2007. Shifts in positive and negative plant interactions along a grazing intensity gradient. *Ecology* 88:188-199.

Graff, P. and M. R. Aguiar. 2011. Testing the role of biotic stress gradient hypothesis. Processes and patterns in arid rangelands. *Oikos* 120:1023-1030.

Gurevitch, J., L. L. Morrow, A. Wallace, and J. S. Walsh. 1992. A metaanalysis of competition in field experiments. *American Naturalist* 140:539-572.

Gurevitch, J, and L. V. Hedges. 2001. Meta-analysis: combining the results of independent experiments. Pages 347-369 *in* S. M. Scheiner and J. Gurevitch, editors. *Design and Analysis of Ecological Experiments.* Second Edition. Oxford University Press Inc., New York, New York.

Hall, R. O. Jr., M. F. Dybdahl, and M. C. VanderLoop. 2006. Extremely high secondary production of introduced snails in rivers. *Ecological Applications* 16:1121-1131.

Hillebrand, H., B. Worm, and H. K. Lotze. 2000. Marine microbenthic community structure regulated by nitrogen loading and grazing pressure. *Marine Ecology- Progress Series* 204:27-38.

Hillebrand, H., and M. Kahlert. 2001. Effect of grazing and nutrient supply on periphyton biomass and nutrient stoichiometry in habitats of different productivity. *Limnology and Oceanography* 46:1881-1898.

Hillebrand, H., G. de Montpellier, and A. Liess. 2004. Effects of macrograzers and light on periphyton stoichiometry. *Oikos* 106:93-104.

Hunter, R. D., and W. D. Russell-Hunter. 1983. Bioenergetic and community changes in intertidal aufwuchs grazed by *Littorina littorea*. *Ecology* 64:761-769.

Irving, A. D. and M. D. Bertness. 2009. Trait-dependent modification of facilitation on cobble beaches. *Ecology* 90:3042-3050.

Kawai, T., and M. Tokeshi. 2004. Variable modes of facilitation in the upper intertidal: goose barnacles and mussels. *Marine Ecology Progress Series* 272:203-213.

Kawai, T., and M. Tokeshi. 2007. Testing the facilitation-competition paradigm under the stress-gradient hypothesis: decoupling multiple stress factors. *Proceedings of the Royal Society* B 274:2503-2508.

Keppel, G., and T. D. Wickens. 2004. *Design and analysis: a researcher's handbook.* Fourth Edition. Prentice Hall.

Krist, A. C. and C. M. Lively. 1998. Experimental exposure of juvenile snails (*Valvata humeralis*) to infection by trematode larvae (Microphallus sp.): infectivity, fecundity compensation and growth. *Oecologia,* 116:575-582.

Krist, A. C., J. Jokela, J. Wiehn, and C. M. Lively. 2004. Effects of host condition on susceptibility to infection, parasite development rate, and parasite transmission in a snail-trematode interaction. *Journal of Evolutionary Biology* 17:33-40.

Liess, A., and H. Hillebrand. 2005. Stoichiometric variation in C:N, C:P, and N:P ratios of littoral benthic invertebrates. *Journal of the North American Benthological Society* 24:256-269.

Liess, A., and H. Hillebrand. 2006. Role of nutrient supply in grazer-periphyton interactions: reciprocal influences of periphyton and grazer nutrient stoichiometry. *Journal of the North American Benthological Society* 25:632-642.

Liess, A., and M. Kahlert. 2007. Gastropod grazers and nutrients, but not light, interact in determining periphytic algal diversity. *Oecologia* 152:101-111.

Liess, A., and K. Lange. 2011. The snail *Potamopyrgus antipodarum* grows faster and is more active in the shade, independent of food quality. *Oecologia* 167:85-96.

López-Figueroa, F., and F. X. Niell. 1987. Feeding behavior of *Hydrobia ulvae* (Pennant) in microcosms. *Journal of Experimental Marine Biology and Ecology* 114:153-167.

McNaughton, S. J. 1976. Serengeti migratory wildebeest: facilitation of energy flow by grazing. *Science* 191:92-94.

Price, J. E. and P. J. Morin. 2004. Colonization history determines alternative community states in a food web of intraguild predators. *Ecology* 85:1017-1028.

Rosemond, A. D. 1993. Interactions among irradiance, nutrients, and herbivores constrain a stream algal community. *Oecologia* 94:585-594.

Ruess, R. W., and S. J. McNaughton. 1984. Urea as a promotive coupler of plant-herbivore interactions. *Oecologia* 63:331-337.

Schoener, T. W. 1983. Field experiments on interspecific competition. *The American Naturalist* 122:240-284.

Smit, C., C. Vandenberghe, J. den Ouden, and H. Muller-Scharer. 2007. Nurse plants, tree saplings and grazing pressure: changes in facilitation along a biotic environmental gradient. *Oecologia* 152:265-273.

Smit, C., M. Rietkerk, and M. J. Wassen. 2009. Inclusion of biotic stress (consumer pressure) alters predictions from the stress gradient hypothesis. *Journal of Ecology* 97:1215-1219.

Sommer, U. 1992. Phosphorus-limited *Daphnia*: intraspecific facilitation instead of competition. *Limnology and Oceanography* 37:966-973.

Stachowicz, J. J. 2001. Mutualism, facilitation, and the structure of ecological communities. *BioScience* 51:235-246.

Tibbets, T. M., A. C. Krist, R. O. Hall, and L. A. Riley. 2009. Phosphorus-mediated changes in life history traits of the invasive New Zealand mudsnail (*Potamopyrgus antipodarum*). *Oecologia* 163:549-559.

Tielborger, K., and R. Kadmon. 2000. Temporal environmental variation tips the balance between facilitation and interference in desert plants. *Ecology* 81:1544-1553.

Underwood, A. J. 1997. *Experiments in Ecology: Their Logical Design and Interpretation Using Analysis of Variance*. Cambridge University Press, Cambridge, UK.

Vanni, M. J., A. S. Flecker, J. M. Hood, and J. L. Headworth. 2002. Stoichiometry of nutrient recycling by vertebrates in a tropical stream: linking species identity and ecosystem processes. *Ecology Letters* 5:258-293.

Vesey-Fitzgerald, D. F. 1960. Grazing succession among East African game animals. *Journal of Mammalogy* 41:161-172.

Von Ende, C. N. 2001. Repeated-measures analysis: growth and other time dependent measures. Pages 134-157 *in* S. M. Scheiner and J. Gurevitch, editors. *Design and analysis of ecological experiments.* Second Edition. Oxford University Press Inc, New York, New York.

Went, F. W. 1942. The dependence of certain annual plants on shrubs in southern California deserts. *Bulletin of the Torrey Botanical Club* 69:100-114.

Whiles, M. R., A. D. Huryn, B. W. Taylor, and J. D. Reeve. 2009. Influence of handling stress and fasting on estimates of ammonium excretion by tadpoles and fish: recommendations for designing excretion experiments. *Limnology and Oceanography: Methods* 7:1-7.

Chapter 5

The Role of Superoxide Dismutase in Reptiles under Toxicity Contexts

L. M. Odetti[1,2,*], E. V. Paravani[3], M. F. Simoniello[1] and G. L. Poletta[1,2,4]

[1]Cátedra de Toxicología, Farmacología y Bioquímica Legal, Facultad de Bioquímica y Ciencias Biológicas, Universidad Nacional del Litoral, Santa Fe, Argentina
[2]Consejo Nacional de Investigaciones Científicas y Técnicas (CONICET), Argentina
[3]Laboratorio de Química Ambiental, Cátedra de Química General e Inorgánica, Facultad de Ingeniería, Universidad Nacional de Entre Ríos, Entre Ríos, Argentina
[4]Proyecto Yacaré-Laboratorio de Zoología Aplicada: Anexo Vertebrados (FHUC-UNL/Gob. Sta. Fe/MAyCC), Santa Fe, Argentina

Abstract

Reptiles are important models in evolutionary history and several species are of great ecological, sociological and commercial importance. They are transcendental in natural ecosystems, as prey, predators, herbivores, seed dispersers and as bioindicators of environmental health. In recent decades, the role of environmental pollution as a possible reason for the decline of reptile populations has prompted researchers to evaluate the effects of pesticides on these vertebrate groups. Given the expansion of agricultural frontiers, many reptile populations live now in habitat subjected to high pollution pressure, and due to their biological characteristics, they may be exposed to pollutants at all stages of their lives. As a consequence, they appear worldwide as excellent model species of environmental contamination by different xenobiotics. Environmental pollutants are known to induce oxidative damage to biomolecules, such as proteins, lipids, and nucleic acids, due to the

[*] Corresponding Author's Email: luodetti@gmail.com.

In: Advances in Animal Science and Zoology. Volume 20
Editor: Owen P. Jenkins
ISBN: 979-8-88697-199-6

production of reactive oxygen species (ROS). ROS are highly reactive and potentially deleterious to biological systems. To minimize the effects that ROS can generate, aerobic organisms have developed enzymatic and non-enzymatic antioxidant defenses. Enzymatic defenses, such as superoxide dismutase (SOD), catalase (CAT) and glutathione peroxidase (GPx), protect the organism by directly eliminating superoxide radicals and hydrogen peroxide, converting them into less reactive species. Oxidative stress occurs as a result of an imbalance between ROS and antioxidant mechanisms, leading to a variety of physiological and biochemical changes that cause cell deterioration and death. This occurs due to an excessive accumulation of ROS, decrease in antioxidant enzymes, or both. Specifically, SOD is the first line of defense and the most powerful antioxidant in the cell. It catalyzes the dismutation of superoxide anion ($O2^{\bullet-}$) into hydrogen peroxide (H_2O_2) and oxygen (O_2), making potentially harmful $O2^{\bullet-}$ less dangerous. Four isoforms of SOD enzyme have been described, according to their metallic cofactor; Fe-SOD, Mn-SOD, Ni-SOD and Cu/Zn-SOD, which have distinct subcellular locations. SOD has demonstrated to be sensible to environmental contamination and several studies carried out in reptiles have shown that exposure to xenobiotics produces inhibition or overproduction of SOD or altered *sod* gene expression in different tissues. Considering the importance of reptiles in the ecosystem and the action of SOD against environmental pollutants, the aim of this chapter is to present the current knowledge into the structure-activity relationship and mechanism of action of SOD in a reptile species, and its importance as a biomarker for environmental contamination in sentinel species.

Keywords: reptiles, sentinel species, environmental pollution, oxidative stress, biomarkers

Brief Introduction

In recent decades, environmental pollution has become a global challenge due to its exponential increase worldwide. It is linked to unsustainable anthropogenic activities, which has a particularly high impact on wildlife (Hooman, 2020). Currently, biodiversity decline is a global problem among all taxa and ecosystems. The causes of these declines are varied, and environmental pollution is believed to be one of the main factors (Todd et al., 2010; Ceballos et al., 2015). Research is providing baseline information for several animal groups, generally due to their sensitivity to changes in their environment, but also where there is a potential conservation threat (Ortiz-

Santaliestra et al., 2018). Specifically, reptiles are one of the most threatened vertebrate groups (Amaral et al., 2012 a, b; Schaumburg et al., 2016; Carpenter et al., 2016; Poletta et al., 2016; Burella et al., 2018; Odetti et al., 2020). According to the International Union for Conservation of Nature (IUCN), these vertebrates are suffering a notable decline in their populations on a global scale. Of all reptile species assessed by the IUCN, about 21% are classified as endangered or vulnerable to extinction (IUCN, 2020). Especially the combination of habitat loss and environmental pollution (mainly pesticides) are among the main causes contributing to the loss of local and regional biodiversity of this group (Todd et al., 2010).

In recent decades, the role of environmental pollution on reptile populations has prompted several investigations to assess the effects of chemicals on this vertebrate groups. In view of their biocidal activity, scale and volume of application, pesticide contamination is currently recognized as one of the main factors threatening reptile populations (Sánchez-Bayo, 2011; Sasaki et al., 2015).

Reptiles as Sentinel Organisms

Reptiles are non-target organisms for enviroment contamination (Todd et al., 2010). They are exposed to them through ingestion of contaminated food, through soil, inhalation, dermal exposure, maternal transfer to eggs or hatchlings, absorption of contaminants by eggs from surrounding environments and as a result of decreased food availability or vegetation cover (Sanchez-Bayo, 2011).

Through the years, reptiles have proven to be good bioindicators of environmental health, revealing early the presence and toxicity of one or more pollutants in the environment. Xenobiotics can affect organisms directly through toxicity and indirectly through changes in biotic and abiotic factors, however, how these mechanisms interact to affect wild reptile populations remains poorly understood (Mishra et al., 2019).

When a xenobiotic is introduced into an environment, its first effects are observed at the level of the organism, then the population, the community and ultimately at the level of the entire ecosystem. The earlier these disturbances are detected, the greater the chances of prevention or remediation and the opportunity to detect the source of contaminants (Berthet et al., 2013; Figure 1).

Particularly, pesticides cause changes at various levels of biological organization (molecular, cellular, physiological, individual, etc.). Therefore, although they remain present in the environment, reptiles (as bioindicators) may reveal changes in their health status or behavioral modifications that could influence their reproductive capacity and their long-term survival in the environment they inhabit, which could lead to altered ecosystem health, structure, and functioning (Kier, 2015). Diferents studies have shown that reptiles are sentinel organisms due to some important characteristics: they are integral components of ecosystems, they are long-lived organisms with fairly restricted home ranges, generally at the top level of food chains and genetically stable, with high and constant sensitivity to toxicants and/or environmental changes (White, 2004; Cunha and Guilhermino, 2006).

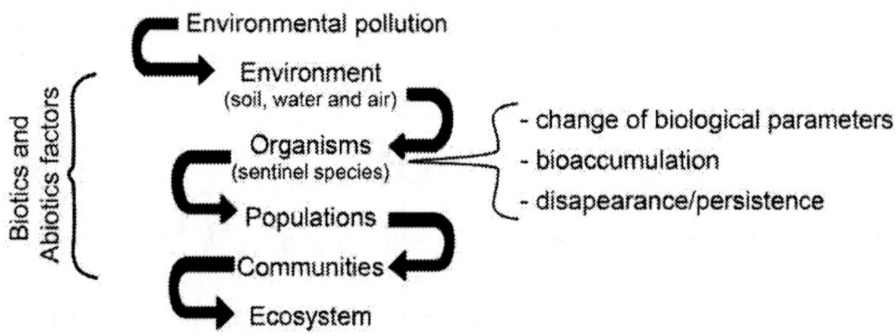

Figure 1. Relationship between sentinel species and the various levels of biological organization.

Over the years, several biomarkers have been developed and implemented to study the mechanisms that could explain the effects induced by pesticides or other environmental pollutants in reptiles. Biomarkers are a crucial tool, because they can be used as early warning signals, before the onset of irreversible damage at the population level (Roméo and Giambérini, 2013). Assessing the impacts of contaminants on the health of organisms and ecosystems is challenging due to the presence of multiple stressors and the complexity of ecosystems. Biomarkers are widely used to provide the connection between external levels of contaminant exposure, internal levels of tissue contamination and early adverse effects in organisms (Carballo and Mudry, 2006). Therefore, the use of biomarkers may be a critical line of evidence for understanding the relationships between stressors and effects on different organisms, and for preventing detrimental impacts of pollution on ecosystem structure and function (Adams, 2004).

According to Carballo and Mudry (2006) for the choice of a biomarker, beside the use of a proper sentinel species, the following aspects should be taken into account: specificity, sensitivity, relevance, feasibility, interspecific and seasonal variation. In turn, biomarkers can be of three types: exposure, effect or susceptibility.

The biomarkers of effect include biochemical, physiological, behavioral or other alterations suffered by the exposed organism, which may or may not be irreversible. Among them, we will pay particular attention to oxidative stress biomarkers in the context of toxicity (Carballo and Mudry, 2006).

Oxidative Stress and Antioxidant Enzymes

Oxygen (O_2) is an essential element for the survival of aerobic organisms and is used by cells for the oxidation of organic compounds, cellular respiration, energy production, among others. However, in all these cellular processes oxygen generates free radicals or reactive oxygen species (ROS), which can damage the cell components. ROS are produced in the complete reduction of O_2 to water (Figure 2). The products of these reactions are superoxide anion radical ($O_2^{\bullet-}$), hydrogen peroxide (H_2O_2) and hydroxyl radical ($OH^{\bullet}$) (Scandalios, 2005).

Both $O_2^{\bullet-}$ and OH^- are free radicals and are defined as chemical species that have one or more unpaired electrons in their outermost orbital. For a chemical compound to become a free radical, it is necessary to gain or lose electrons. $O_2^{\bullet-}$ and $OH^{\bullet}$ are highly reactive and can cause severe molecular damage. H_2O_2 is a reducing agent and a weak oxidant. Its neutral nature allows it to penetrate cell membranes. Its major importance in oxidative stress (OS) is related to its ability to serve as a precursor of the $OH^{\bullet}$ radical (Halliwell,2006).

ROS are highly reactive and potentially deleterious to biological systems. To minimize the effects that ROS can generate, aerobic organisms have developed enzymatic and non-enzymatic antioxidant defenses. Non-enzymatic defenses include vitamins A, C and E, glutathione, ferritin, among others. Enzymatic defenses such as suporexide dismutase (SOD), catalase (CAT) and glutathione peroxidase (GPx), protect cells by directly scavenging superoxide radicals and hydrogen peroxide, converting them into less reactive species (Ighodaroa and Akinloyeb, 2018).

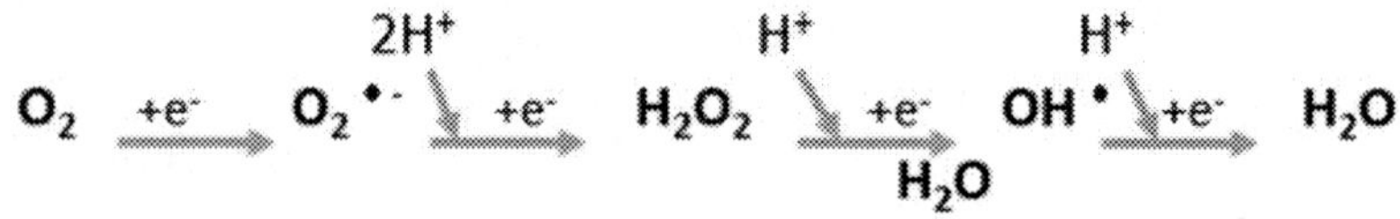

Figure 2. Pathways in the reduction of oxygen in water leading to the generation of ROS.

SOD is the first detoxification enzyme and the most powerful antioxidant in the cell. It catalyzes the dismutation of two $O_2^{\bullet -}$ molecules into H_2O_2 and O_2, which makes potentially harmful $O_2^{\bullet -}$ less dangerous. CAT catalyzes the degradation or reduction of H_2O_2 into water and O_2, thus completing the detoxification process. SOD and CAT are the first line of antioxidant defense (Halliwell, 2006). GPx, in turn, metabolizes a variety of peroxides, including H_2O_2. Among the non-enzymatic antioxidants is the tripeptide glutathione in its reduced form (GSH), which acts as the main antioxidant in the cell and is a cofactor for the action of glutathione-S-transferase (GST) and GPx. GST is an enzyme that acts in the biotransformation process, catalyzing the conjugation of a variety of metabolites, including xenobiotic metabolites and lipoperoxidation products with GSH, transforming the toxic compound into a more easily excretable one (Birben et al., 2012).

ROS such as $O_2^{\bullet -}$, H_2O_2 and $OH^{\bullet}$ are produced in all aerobic organisms and normally exist in cells in equilibrium with the antioxidant system (AS). OS occurs as results of an imbalance between ROS and antioxidant mechanisms, leading to a variety of physiological and biochemical changes that cause cell deterioration and death. This occurs due to a decrease in antioxidant enzymes or excessive accumulation of ROS, or both (Adams et al., 2015; Figure 3). OS can cause damage to many components of the cell, including proteins, lipids, DNA and RNA (Ighodaroa and Akinloyeb, 2018).

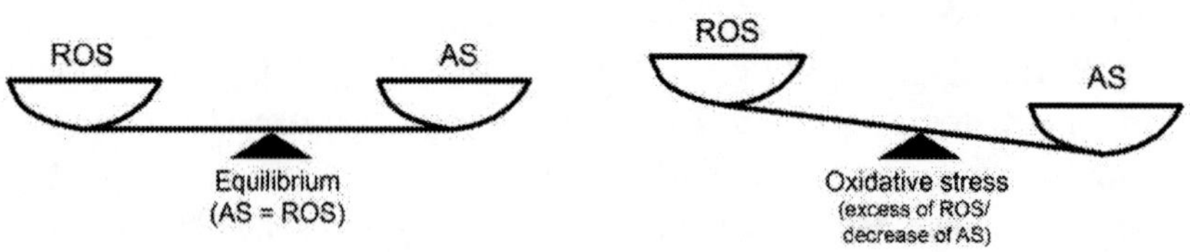

Figure 3. Cells can normally balance the production of oxidants and antioxidants to maintain redox balance. OS occurs when this balance is disturbed by excessive levels of ROS or decreased antioxidant defenses.

Environmental pollutants, and especially pesticides, can act as potent inducers of OS (Mingo et al., 2016). In this sense, they act as prooxidants in a variety of tissues; producing accumulation of ROS, DNA damage, alteration of antioxidant defenses and LPO, causing major perturbations to intra- and intercellular homeostasis (Go and Jones, 2008; Birben et al., 2012; Ighodaroa and Akinloyeb, 2018). A large number of studies carried out in mammals, amphibians and fish have shown that exposure to these compounds produces ROS and oxidative damage, thus generating deficiencies in exposed organisms (Rui et al., 2012; Mu et al., 2013; Ullah and Zorrieh-Zahra, 2015).

In the following, particular attention will be paid to the role of superoxide dismutase in reptiles in the context of toxicity.

Superoxide Dismutase in Reptiles

SODs are antioxidant metalloenzymes and there are four isoforms according to their metallic cofactor; Iron (Fe-SOD), Manganese (Mn-SOD), Nickel (Ni-SOD) and Copper/Zinc (Cu/Zn-SOD), which have distinct subcellular locations (Bafana et al., 2012). Cu/Zn-SOD is found in intracellular and extracellular compartments of both prokaryote and eukaryote, Mn-SOD in mitochondria and peroxisome, Fe-SOD in chloroplast, and Ni-SOD is exclusive of Streptomyces (Duarte et al., 2010). Cu, Zn-SOD represents more than 90% of the total SOD of an organims and requires Cu and Zn for its biological activity. The loss of Cu results in its complete inactivation and induces many diseases in human and animals (Bafana et al., 2012).

The activity of SOD has demonstrated to be sensible to environmental contamination (Paravani et al., 2020). However, variations in SOD activity may depend on the characteristics of each study, as inhibition, activation or both have been reported in tissues from different organisms (Weidinger and Kozlov, 2015). Besides, different investigations have shown that the analysis of gene expression patterns or the enzymatic activity of SOD, could help to improve our understanding of the early effects of environmental contamination (Aguilera et al., 2012; Héritier et al., 2017a; Burella et al., 2018; Paravani et al., 2020).

Studies on the effects of exposure to environmental contaminants on reptiles are limited, and refer mainly to the orders Testudines and Crocodylia, but largely neglect the more diverse reptile order Squamata. Consequently, information regarding the effects on these organisms as well as the likelihood of exposure and uptake of environmental contaminants in reptiles should be considered unexplored (Mingo et al., 2017).

Studies in Reptile Species Worldwide

Reptiles are important models in evolutionary history, and they have great ecological, sociological and commercial importance (Sasaki et al., 2015). In ecotoxicology, reptiles are the least studied group of animals even when several species are declining worldwide, mainly due to the degradation of their habitat by anthropogenic action (Todd et al., 2010). Valdivia et al. (2007) aimed obtaining baseline information of oxidative stress indicators in eastern Pacific green turtles (*Chelonia mydas agassizii*) from a relatively undisturbed habitat at Bahía Magdalena, Baja California Sur, Mexico. In this work, the activity of the antioxidant enzyme SOD was analyzed in heart, liver, lung, kidney and pectoral muscle. In general, the results suggest differences in antioxidant system among tissues. Other authors selected juveniles of the freshwater *Trachemys scripta elegans* as a model to measure enzyme activity and *sod* gene expression in turtles exposed to high concentrations of glyphosate (GLY) and contaminated water from a local watercourse in France, showing statistically significant increases in exposed animals over the controls (Héritier et al., 2017a). Same authors determined the activity of SOD in blood of the endangered freshwater endemic species *Mauremys leprosa* and standardized the quantitative real-time PCR (qPCR) technique for the analysis of gene expression (*sod*) to conceive environmental ecotoxicological studies without euthanizing animals. Then, Héritier et al. (2017b), investigated the effects of degraded waters on the physiology of the same species, in contrasting environments along the same rivers in the Eastern Pyrenees (France). In this work, they developed a procedure to study the activity of SOD and CAT and the expression level of their correspondinig genes in blood samples, finding significant variations in gene expression and CAT activity upstream or downstream of wastewater treatment plants.

On the other hand, Salvarani et al. (2018) evaluated the health status of sea turtles (*Eretmochelys imbricata*) and the effect of organochlorine compounds in the southern Gulf of Mexico. They looked for relationships between carapace size and antioxidant enzyme activity in hawksbill turtle blood at Punta Xen (Campeche, Mexico). The results of this study provide data on antioxidant enzyme activities in relation to organochlorine contaminants and their possible impacts in sea turtles health.

Crocodiles, in particular because of their position in the food chain, their aquatic habits and their longevity, may reflect changes in an area they inhabit over longer periods (Todd et al., 2010). Specifically, in *Alligator mississippiensis* it was investigated the potential of using tail scales as nondestructive (not requiring euthanasia) indicators of the antioxidant enzyme SOD. The results obtained suggest that muscle (from tail scale) antioxidant enzyme levels may serve as important nondestructive indicators of the oxidative status of internal organs and the whole body. In addition, antioxidants in tail scales may also provide information on the body condition and health of alligators (John et al., 2021). For crocodilian species, there are only reports on nucleotide sequences of the Cu, Zn-*sod* gene for *Alligator mississippiensis*, *Alligator sinensis* and *Crocodylus siamensis*, derived from the crocodile genome sequencing project (Wan et al., 2013; Green et al., 2014; Sujiwattanarat et al., 2016). Specifically, Sujiwattanarat et al. (2016) was the first report about molecular cloning and characterization of Siamese crocodile (*Cr. siamensis*) copper, zinc- superoxide dismutase gene (Cu, Zn-*sod*), and the analysis of its expression levels in different tissues (liver, pancreas, lung, kidney, heart and whole blood) by qPCR.

In terrestrial environments, lizards are particularly suitable as pollution biomonitors, as a result of their presence in diverse habitats, their wide geographic distribution, longevity and, in many cases, site fidelity (Croteau et al., 2008). In addition, lizards are especially useful during dry seasons or in arid regions (Araujo,2020). In this context, Aguilera et al., (2012) determined the levels of enzymatic activities, including SOD in wild populations of spiny lizards (*Sceloporus serrifer* and *S. torquatus*) from two different localities near the metropolitan area of Monterrey, Mexico: Chipinque Ecological Park, a natural protected area, and El Carmen Industrial Park, a highly polluted site. The investigation was based on the use of tail tips, which regenerate easily, as a non-destructive technique. The activity of SOD was statistically higher in the exposed area with respect to the control area.

In the same way Chen et al. (2019) studied the toxic effects of beta-cypermethrin and myclobutanil in saliva, serum, liver, kidney, brain, and gonad in *Eremias argus*. In this work, a single dose of both compounds was orally administered at the concentration of 20 mg/kg body weight, and lizards were euthanized at 96 h. The results showed that pesticides could induce changes in enzymatic activity of SOD in organs or tissues of the lizards. Particulary, salivary enzymes activities showed sensitivity to the toxicity of pesticides and could be used as an interesting and low invasive sample.

Table 1 summarizes the reported studies in reptiles under different conditions of exposure and includes the tissue and different life stages.

Table 1. Oxidative stress studies in reptiles from different parts of the world

Species	Assessment conditions	Determinations	Number of animals	Age	Tissue used	Reference
Chelonia mydas agassizii	Baseline values of oxidative stress biomarkers in eastern Pacific	-Enzymatic activities (SOD, CAT and GST) -Lipoperoxidation	19	Juvenile	Heart, liver, lung, kidney and pectoral muscle	Valdivia et al., 2007
Sceloporus serrifer and *S. torquatus*	*In situ* study of environmental pollution from two different locations near the metropolitan area of Mexico	-Enzymatic activities (SOD and GST)	27	Juvenile	Tail tips	Aguilera et al., 2012
Trachemys scripta elegans; Mauremys leprosa	a) Glyphosate and contaminated water from a local watercourse in France: short-term. b) Standardization of qPCR technique	-Enzymatic activities (SOD and CAT) -Gene expression (*sod* and *cat*) -Lipoperoxidation	41	Juvenile	Liver and blood	Héritier et al., 2017a
Mauremys leprosa	*In situ* study in relation to environmental pollution in the Pyréenées-Orientales department, France	-Enzymatic activities (SOD and CAT) -Gene expression (*cat* and *sod*)	163	Juvenile	Blood	Héritier et al., 2017b
Eretmochelys imbricata	*In situ* study in relation to environmental pollution in the southern part of the Gulf of Mexico.	-Enzymatic activities (SOD, CAT, GR and GST)	60	Adult	Blood	Salvarani et al., 2018
Eremias argus	Beta-cypermethrin and myclobutanil: short- term	-Enzymatic activities (SOD and CAT) -Lipoperoxidation	18	Juvenile	Saliva, brain, liver, kidney, testes	Chen et al., 2019
Alligator mississippiensis	Baseline values in scutes as potential non-invasive indicators of oxidative status	-Enzymatic activities (SOD and GP_X)	-	Juvenile	Scutes	John et al., 2021
Chelonia mydas	*In situ* study in relation to environmental contamination in the Gulf of Guinea, Sao˜ Tomé (Africa)	-Gene expression (*sod, cat, gr, tdx, txrd, selp* and *gclc*)	27	Adult	Blood	Morao et al., 2022

Enzymatic activities: suporexide dismutase (SOD), catalase (CAT), glutathione peroxidase (GPx), glutathione reductase (GR) and glutathione S-transferase (GST). Gene expression: suporexide dismutase (*sod*), catalase (*cat*), glutathione reductase (*gr*), glutathione S-transferase (*gst*), thioredoxin (*tdx*), thioredoxin reductase (*txrd*), selectin P (*selp*) and glutamate-cysteine ligase (*gclc*).

Studies in Reptile Species from South America

Oxidative stress studies related to South American species are quite scarce and most of them belong to our group (Poletta et al., 2016; Burella et al., 2018; Paravani et al., 2020; Odetti et al., 2020, 2021; López-Gonzaléz et al., 2021).

One of the lizards with a wide distribution in South America is *Salvator merianae* (tegu lizard). It is frequently found in areas of agricultural crops, and its reproductive period coincides with the moment of maximum application of pesticide in spring and summer (Croteau et al., 2008). For this species, Araujo (2020) studied the possible changes in the biochemical parameters in animals from an area exposed to agrochemicals compared to a reference area and determined SOD activity in blood. This work concluded that contact with pesticides is causing a deleterious effect on the activity of SOD in *S. merianae*, and that it may have negative consequences in this population.

Caiman latirostris (broad snouted caiman) is one of the two species of crocodilians that inhabits the north-central region of Argentina, and it constitutes a valuable commercial and ecological resource in South America. The geographical distribution of this species includes the northeaster of Argentina, southeaster Bolivia, Paraguay, Brazil, and northern Uruguay. Particularly in Argentina, *C. latirostris* is distributed in the provinces of Formosa, Santa Fe, Misiones, Corrientes, Entre Ríos, Chaco, Santiago del Estero, Salta, and Jujuy (Siroski et al., 2020).

One of the main problems affecting their wild populations during the last 25 years is habitat loss due to anthropogenic pressure, tending to assign more and more lands to agricultural production (Poletta et al., 2009). In this context, Poletta et al., (2016) developed a new set of biomarkers of oxidative stress in blood of this species through the modification of conventional techniques, including the determination of the activity of the antioxidant enzyme SOD. This was the first report of the application of these biomarkers in blood for all crocodilians species and allowed to apply a complete battery of tests without any damage to the animals. Then, Burella et al. (2018) analyzed the activity of SOD enzyme in erythrocytes of caimans exposed to sublethal concentrations of two GLY formulations: PanzerGold® (PANZ) and Roundup® (RU) Full II (500–1000 μg/egg); endosulfan (END) formulation Galgofan® and Cyphermetrin (CYP) formulation Atanor® (1–1000 μg/egg) by topical application through the eggshell at the beginning of the incubation period, showing an induction in groups exposed to the insecticides. Recently, alterations in the activity of SOD in erythrocytes was demonstrated in

C. latirostris hatchlings (20 days old) exposed *in vivo* to sublethal concentrations of PANZ and the ternary mixture of RU + END + CYP. Animals exposed to this mixture showed besides, a significantly higher SOD activity than that induced separately by two of the three constituting compounds (RU and CYP (López-Gónzalez et al., 2021).

Considering these results and in order to better understand the role of this enzyme in the prevention of oxidative stress, we determined the presence, characterized the molecular structure and quantified the tissue-specific expression of Cu, Zn-*sod* in *C. latirostris*, including liver, gonads, spleen, heart and whole blood through a PCR (Paravani et al., 2020). Figure 4 shows one of the best quality PCR products obtained.

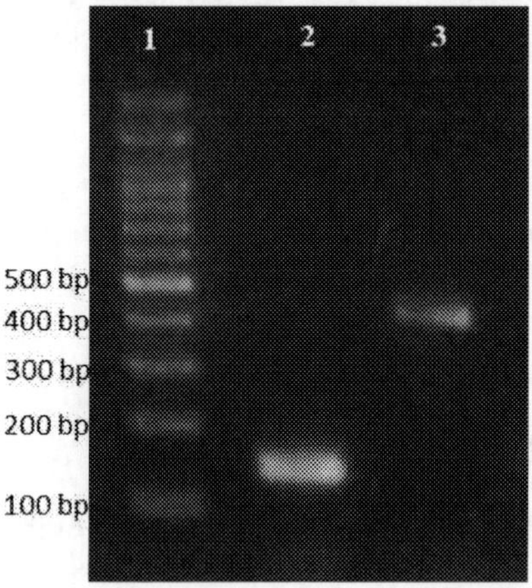

Figure 4. PCR products obtained from *C. latirostris* blood. Agarose gel (2%) SYBR® Safe (Invitrogen) stained. Lane 1: Ladder 100 pb (Plus DNA - TransGen Biotech). Lane 2: *C. latirostris* β-actin (138 pb). Lane 3: *C. latirostris* Cu/Zn-*sod* for primer pair 3 (444 pb). Paravani et al., (2020).

In turn, in this work, we utilized the homology modeling tools to visualize the three-dimensional structure of *C. latirostris* Cu/Zn-SOD. The majority of Cu/Zn-SOD are homodimeric proteins, in which the monomeric subunits are stabilized by a disulfide bond, close to the active site (Tainer et al., 1982). The predicted global structure of *C. latirostris* Cu/Zn-SOD was comparable with the tertiary structure of Cu/Zn-SOD, where there are antiparallel β-strands, turns or short α-helices, in addition to the active sites characterized by Cu and Zn ions (Figure 5).

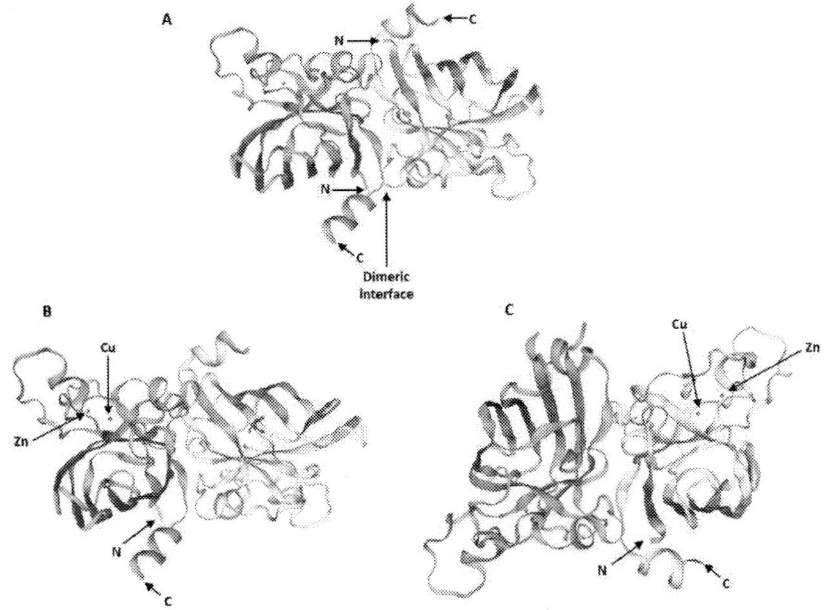

Figure 5. Homology modeling of *C. latirostris* Cu/Zn-SOD tertiary structure. (A) Semi-developed 3D model representation of homodimeric structure of Cu/Zn-SOD, based on homo-dimer crystal structure of human extracellular Cu/Zn-SOD. Dimeric interface is indicated by a black arrow. The C- and N-terminals are indicated by respective letters. (B-C) Topology copper-binding residues and the binding pocket for zinc. Paravani et al. (2020).

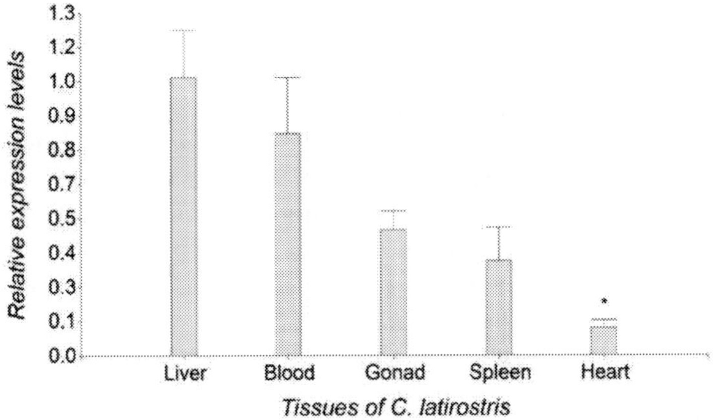

Figure 6. Semi-quantitative gene expression levels of Cu/Zn-*sod* in different tissues of subadult *C. latirostris* (n = 5). Paravani et al. (2020).

Then, the primer designed from the consensus sequence obtained for *C. latirostris* Cu/Zn-*sod* allowed the determination of gene expression in different tissues (liver, blood, spleen, gonads and heart) with an amplicon of 442 bp. Results of the semi-quantitative analysis of Cu/Zn-*sod* gene are presented in Figure 6. Considering gene expression level, there was only a significant difference between heart and liver ($p = 0.025$, Kruskal Wallis – Dunn-Bonferroni post hoc test). The liver and blood were the two tissues with higher Cu/Zn-s*od* gene expression, while the heart showed the lowest level. No significant differences were observed among the other tissues.

Recently, we standardized the qPCR technique for quantitative analysis of genes involved in the main routes of antioxidant defense (*cat* and *sod*) against reactive oxygen species and evaluated the stability of different reference genes in blood of *C. latirostris* hatchlings. This was the first report of the analysis of antioxidant mRNA as potential biomarkers of oxidative stress in blood for all crocodilians species (Odetti et al., 2021). In order to use gene expression levels as potential biomarkers of pesticide effects, we evaluated the embryonic exposure to GLY, CYP and CPF-based formulations separately and in mixtures in *C. latirostris* blood, simulating the exposure conditions that can occur in natural environments through recommended concentrations and schedule for application in soybean crops. These conditions generate significant under-expression of *sod* genes in the mixtures tested, demonstrating that this is a sensitive marker. This was the first report of gene expression alteration pertaining to the antioxidant system in response to pesticide exposure for all crocodilian species (Odetti et al., under review).

The underexpression of the *sod* gene is possibly due to an inhibition in the regulation of its expression, directly caused by the toxic action of the pesticides. Furthermore, previuously we had found no significant differences in SOD enzymatic activity under the same exposure conditions (Odetti et al., 2020). In general, and contrary to our results, the expression of genes encoding antioxidant enzymes is up-regulated when the cell is challenged by ROS, and is under the control of numerous transcriptional factors activated by ROS (Liu et al., 2015). However, some authors suggest that mRNA levels are representative of cellular activity at a particular time, whereas protein activity might be regulated at the post-translational level, enzymatically and not by transcriptional control (Jin et al., 2010).

Table 2 summarizes SOD studies reported for South American reptile species under different conditions.

Table 2. Oxidative stress studies in reptiles from South America

Species	Assessment conditions	Determinations	Age	Tissue used	Result	Reference
Salvator merianae	*In situ* study of environmental pollution at the Experimental Agronomic Station of UFRGS, Brazil	Enzymatic activities (SOD, CAT and GST) Lipoperoxidation	Juvenile	Blood	+ +	Araujo, 2020
Caiman latirostris	Baseline values of oxidative stress biomarkers	Enzymatic activities (SOD, CAT and GST) Lipoperoxidation Oxidative DNA damage	Juvenile	Blood	na	Poletta et al., 2016
	Embryonic exposure to sub-lethal concentrations of GLY, END and CYP: by topical application on the eggshell	Enzymatic activities (SOD and CAT) Lipoperoxidation Oxidative DNA damage	Embryos	Blood	+ + +	Burella et al., 2018
	Detection and molecularcharacterization of the Cu, Zn-*sod* gene and protein	Presence and gene expression patterns	Subadults	Liver, blood, spleen, gonads and heart	na	Paravani et al., 2020
	Embryonic exposure to environmentally relevant concentrations of CYP, CPF, GLY and mixtures by spraying on the nest material in contact with the eggs	Enzymatic activities (SOD and CAT) Lipoperoxidation Oxidative DNA damage	Embryos	Blood	- - +	Odetti et al., 2020
	Sub-chronic exposure to GLY, END and CYP and mixtures by voluntary immersion in plastic experimental containers	Enzymatic activities (SOD and CAT) Lipoperoxidation	Hatchlings	Blood	+ +	López-Gonzaléz et al., 2021
	Standardization the qPCR technique for the analysis of genes involved in the main routes of antioxidant defense.	Gene expression *cat* and *sod* genes	Hatchlings	Blood	na	Odetti et al., 2021

Table 2. (Continued)

Species	Assessment conditions	Determinations	Age	Tissue used	Result	Reference
Caiman latirostris (Continued)	Semi-natural (*ex situ*) exposure to environmentally relevant concentrations of CYP, CPF, GLY and mixtures	Enzymatic activities (SOD and CAT) Lipoperoxidation Oxidative DNA damage	Juvenile	Blood	- - +	López-Gonzaléz et al., 2022
	Embryonic exposure to environmentally relevant concentrations CYP, CPF, GLY and mixtures by spraying on the nest material in contact with the eggs	Gene expression (*sod* and *cat*)	Embryos	Blood	+	Odetti et al., under review

na: not applicable. Enzymatic activities: suporexide dismutase (SOD), catalase (CAT), and glutathione S-transferase (GST). Gene expression: suporexide dismutase (*sod*), catalase (*cat*). Pesticides: Cypermethrin (CYP), Endosulfan (END), Chlorpyrifos (CPF), glyphosate (GLY).

Conclusion

From the information presented in this chapter, we can conclude that there are few works on this subject and that the measurement of antioxidant enzymatic activity is a strong indicator of the detrimental effects of environmental pollutants on reptile species. The application of new biomarkers, such as the determination of gene expression levels, together with the biochemical parameters and/or cytological or histological abnormalities usually analyzed, will allow us to obtain complementary information, providing an integral perspective of the effects of these compounds in different sentinel species. It is necessary to continue studying the possible metabolic pathways affected by environmental contaminants, in order to identify basic molecular mechanisms that explain the physiological responses observed, and allow us to estimate the long-term consequences for environmentally exposed wild populations.

Acknowledgments

All of the studies presented in this chapter were supported by the Agencia Nacional de Promoción Científica y Tecnológica (PICT 2016-2020 to GLP), Consejo Nacional de Investigaciones Científicas y Técnicas (PIP 11220130100478CO to GLP and Proyecto Yacaré - Yacarés Santafesinos (Gob. Sta. Fe/ MAyCC).

References

Adams, L., Franco, M. C. and Estevez, M. C. (2015). Reactive nitrogen species in cellular signaling. *Experimental Biology and Medicine*, 711-717.

Adams, S. M. (2004). Assessing cause and effect of multiple stressors on marine systems. *Marine Pollution Bulletin*, 649–57.

Aguilera, C., Pamela, G. P., Roberto, M. A., David, L. and Julio, C. (2012). Pollution biomarkers in the spiny lizard (*Sceloporus* spp.) from two suburban populations of Monterrey, Mexico. *Ecotoxicology*, 2103–2112.

Amaral, M. J., Carretero, M. A., Bicho, R. C., Soares, A. M. and Mann, R. M. (2012a). The use of a lacertid lizard as a model for reptile ecotoxicology studies part 1: field demographics and morphology. *Chemosphere*, 757-764.

Amaral, M. J., Bicho, R. C., Carretero, M. A., Sanchez-Hernandez, J. C., Faustino, A. M. R., Soares, A. M. and Mann, R. M. (2012b). The use of a lacertid lizard as a model for reptile ecotoxicology studies part 2: biomarkers of exposure and toxicity among pesticide exposed lizards. *Chemosphere*, 765-774.

Araujo, 2020. Biomarcadores sanguíneos ligados ao metabolismo e ao estresse em salvator merianae: uma relação com o uso de agrotóxicos. *Programa de pós-graduação em ecologia e evolução da biodiversidade.* http://tede2. pucrs.br/tede2/handle/tede/9233.

Bafana, A., Dutt, S., Kumar, A., Ahuja, P. S. (2012). The basic and applied aspects of superoxide dismutase. *Journal of Molecular Catalysis B: Enzymatic*, 129–138.

Berthet, B. 2013. "Sentinel Species." In: *Ecological Biomarkers Indicators of Ecotoxicological Effects*, edited by Amiard-Triquet, C., Amiard, J. C., Rainbow, P. S, 155-186. CRC Press, Boca Ratón, EUA.

Birben, E., Sahiner, U. M., Sackesen, C., Erzurum, S. and Kalayci, O. (2012). Oxidative Stress and Antioxidant Defens. *World Allergy Organization Journal*, 9-19.

Burella, P. M., Odetti, L. M., Simoniello, M. F. and Poletta, G. L. (2018). Oxidative damage and antioxidant defense in *Caiman latirostris* (Broad snouted caiman) exposed in ovo to pesticide formulations. *Ecotoxicological Environmental Safety*, 437-443.

Carballo, A., Mudry, M. D. 2006. "Indicadores y marcadores biológicos." In: *Genética Toxicológica*, edited by Mudry, M. D., Carballo, M. A, 83-108. De los Cuatro Vientos Editorial, Buenos Aires.

Carpenter, J. K., Monks, J. M. and Nelson, N. (2016). The effect of two glyphosate formulations on a small, diurnal lizard (*Oligosoma polychroma*). *Ecotoxicology*, 548-554.

Ceballos, G., Ehrlich, P. R., Barnosky, A. D., García, A., Pringle, R. M. and Palmer, T. M. (2015). Accelerated modern human–induced species losses: Entering the sixth mass extinction. *Science Advances*, 1:e1400253.

Chen, L., Jinling, D., Wenjun, Z., Luyao, Z., Zikang W., Yao Li Y. D. and Zhiqiang Z. (2019). Effects of beta-cypermethrin and myclobutanil on some enzymes and changes of biomarkers between internal tissues and saliva in reptiles (*Eremias argus*). *Chemosphere*, 69-74.

Croteau, Maxine C., Natacha H., Jennifer C. Gibson, David L., and Vance L. Trudeau. 2008. "Toxicological Threats to Amphibians and Reptiles in Urban Environments." In: *Herpetological Conservation*, edited by Mitchell, J. C., Jung Brown, R. E. and Bartholomew, B, 197–209. The Society for the Study of Amphibians and Reptiles Urban Herpetology.

Cunha, I. and Guilhermino, L. (2006). Environmental Monitoring Protocol for Selection of Sentinel Species and Collection of Specimens. *EROCIPS*, 1-3. In: https://www.ciimar.up.pt/hns/documents/postspill/EROCIPS_-_2006_-_Selection_and_Collection_of_Sentinel_Species_Guideline.pdf Acceso abril 2020.

Duarte, M. M., Moresco, R. N., Duarte, T., Santi, A., Bagatini, M. D., Da Cruz, I. B., Schetinger, M. R. and Loro, V. L. (2010). Oxidative stress in hypercholesterolemia and its association with Ala16Val superoxide dismutase gene polymorphism. *Clinical Biochemistry*, 1118 – 1123.

Go, Y. M., Jones, D. P. (2008). Redox compartmentalization in eukaryotic cells. *Biochimica et biophysica acta*, 1273-1290.

Green, R. E., Braun, E. L., Armstrong. J., Earl. D., Nguyen. N., Hickey. G., Vandewege, M. W., John, J. A., Capella-Gutiérrez, S., Castoe, T. A., Kern, C., Fujita, M. K., Opazo,

J. C., Jurka, J., Kojima, K. K., Caballero, J., Hubley, R. M., Smit, A. F., Platt, R. N., Lavoie, C. A., Ramakodi, M. P., Finger, J. W., Suh, A., Isberg, S. R., Miles, L., Chong, A. Y., Jaratlerdsiri, W., Gongora, J., Moran, C., Iriarte, A., Cormack, J., Burgess, S. C., Edwards, S. V., Lyons, E., Williams, C., Breen, M., Howard, J. T., Gresham, C. R., Peterson, D. G., Schmitz, J., Pollock, D. D., Haussler, D., Triplett, E. W., Zhang, G., Irie, N., Jarvis, E. D., Brochu, C. A., Schmidt, C. J., McCarthy, F. M., Faircloth, B. C., Hoffmann, F. G., Glenn, T. C., Gabaldón, T., Paten, B. and Ray, D. A. (2014). Three crocodilian genomes reveal ancestral patterns of evolution among archosaurs. *Science*, 125-4449.

Halliwell, B. (2006). Reactive species and antioxidants. redox biology is a fundamental theme of aerobic life. *Plant Physiol*, 312-322.

Héritier, L., David, D., Richard, G., Anne-Leila, M. and Olivier, V. (2017a). Oxidative stress induced by glyphosate-based herbicide on freshwater turtles. *Environmental toxicology and chemistry*, 3343–3350.

Héritier, L., Anne-Leila, M. and Olivier, V. (2017b). Oxidative stress biomarkers in the Mediterranean pond turtle (*Mauremys leprosa*) reveal contrasted aquatic environments in Southern France. *Chemosphere*, 332-338.

Hooman, A. (2020). Investigating Energy Use, Environment Pollution, and Economic Growth in Developing Countries. *Environmental and Climate Technologies*, 275–293.

Ighodaroa, O. M. and Akinloyeb, O. A. (2018). First line defence antioxidants-superoxide dismutase (SOD), catalase (CAT) and glutathione peroxidase (GPX): Their fundamental role in the entire antioxidant defence grid. *Alexandria Journal of Medicine*, 287-293.

International Union for Conservation of Nature. (2020). *The IUCN Red List of Threatened Species.* Version 2015-4. http://www.iucnredlist.org.

Jin, Y., Zhang, X., Shu, L., Chen, L., Sun, L., Qian, H., Liu, W. and Fu, Z. (2010). Oxidative stress response and gene expression with atrazine exposure in adult female zebrafish (*Danio rerio*). *Chemosphere*, 846–852.

John, W. F., Meghan, D. K, Yufeng, Z., Matthew, T. H., Ruth, M. E., Mary, T. M. and Andreas, N. K. (2021). Antioxidant Enzymes in Destructible and Non-Destructible Tissues in American Alligators (*Alligator mississippiensis*). *South American J. of Herpetology*, 33-41.

Kier, L. (2015). Review of genotoxicity biomonitoring studies of glyphosate-based formulations. *Toxicology*, 209-218.

Liu, T., Li, L., Zhang, Z. and Wang, Y. (2015). Transcriptional inhibition of the Catalase gene in phosphine-induced oxidative stress in *Drosophila melanogaster*. *Pesticide Biochemistry and Physiology*, 1–7.

López-González, EC., Romito, L. M, Latorre, M. A, Siroski, P. A. and Poletta, G. L. (2021). Biomarkers of genotoxicity, immunotoxicity and oxidative stress on *Caiman latirostris* (Broad-snouted caiman) hatchlings exposed to pesticide formulations and mixtures widely used in agriculture. *Environmental Advances*, 5.

López González, E C., Odetti, L. M., Latorre, M. A., Ávila, O. B., Contini, L. E., Siroski, P. A., Poletta GL. (2022). A comprehensive approach using multiple biomarkers to detect damage induced by pesticides in broad-snouted caiman (*Caiman latirostris*) under ex situ conditions. *Heliyon*.

Mingo, V., Stefan, L. and Norman W. (2017). The impact of land use intensity and associated pesticide applications on fitness and enzymatic activity in reptiles—A field study. *Science of the Total Environment*, 114–124.

Mingo, V., Stefan, L. and Norman, W. (2016). Risk of pesticide exposure for reptile species in the European Union. *Environmental Pollution*, 164-169.

Morao, I.., Lemos, M., Féliz, R., Vieira, S., Barata, C. and Novais, S. (2022). Stress response markers in the blood of Sao Tomé green sea turtles (*Chelonia mydas*) and their relation with accumulated metal levels. *Environmental Pollution*, 293.

Odetti, L. M., López-González, E. C., Romito, M. L., Simoniello, M. F. and Poletta G. L (2020). Genotoxicity and oxidative stress in *Caiman latirostris* hatchlings exposed to pesticide formulations and their mixtures during development through incubation material. *Ecotoxicology and Environmental Safety*, 110-312.

Odetti, L. M., Paravani, E. V., Simoniello, M. F. and Poletta, G. L. (2021). Identification and evaluation of antioxidant and reference genes for quantitative real-time PCR in blood of *Caiman latirostris*. *Heliyon*.

Odetti, LM., Simoniello, M. F. and Poletta, G. L. Alterations in the expression of antioxidant enzyme genes in response to pesticide exposure during embryonic development in the native reptile species *Caiman latirostris*. *Environmental Toxicology and Chemistry*, under review.

Ortiz-Santaliestra, M. E., Joao, P., Maia, A. E. and Isabel, L. (2018). Validity of fish, birds and mammals as surrogates for amphibians and reptiles in pesticide toxicity assessment. *Ecotoxicology*, 819–833.

Paravani, E. V., Odetti, L. M., Simoniello, M. F. and Poletta, G. L. (2020). Molecular analysis and bioinformatic characterization of cooper, zinc-superoxide dismutase (Cu/Zn-*sod*) gene of *Caiman latirostris*. *Molecular Biology Reports*, 8849–8857.

Poletta, G. L., Larriera, A., Kleinsorge, E and Mudry, M. D. (2009) Genotoxicity of the herbicide formulation Roundup® (glyphosate) in broad-snouted caiman (*Caiman latirostris*) evidenced by the Comet assay and the Micronucleus test. *Mutation Research*, 95-102.

Poletta, G. L., Simoniello, M. F. and Mudry, M. D. (2016). Biomarkers of oxidative damage and antioxidant defense capacity in *Caiman latirostris* blood. *Comparative Biochemistry and Physiology Part C: Toxicology & Pharmacology*, 29-36.

Roméo, M. and Giambérini, L. 2013. "History of Biomarkers." In: *Ecological Biomarkers Indicators of Ecotoxicological Effects*, edited by Amiard-Triquet, C., Amiard, J. C., Rainbow, P. S, 15-44. CRC Press, Boca Ratón, EUA.

Rui, P. M., Neuparth, T. S., Caeiroa, S., Lobo, S., Martins, M., Ferreira, A. M., Caetano, M., Vale, C., DelValls, T. A. and Costa, M. H. (2011). Assessment of the genotoxic potential of contaminated estuarine sediments in fish peripheral blood: Laboratory versus in situ studies. *Environmental Research*, 25-36.

Salvarani, P. I., Luis, R., Vieira, W. K., Fernando, M. and Jaime R. O. (2018). Oxidative stress biomarkers and organochlorine pesticides in nesting female hawksbill turtles *Eretmochelys imbricata* from Mexican coast (Punta Xen, Mexico). *Environmental Science and Pollution Research*, 23809–23816.

Sánchez-Bayo, F. 2011. "Impacts of agricultural pesticides on terrestrial ecosystems." In: *Ecological Impacts of Toxic Chemicals*, edited by Sánchez-Bayo, F., Van den Brink, P. J. and Mann, R. M, 75-107. Bentham Publishers.

Sasaki, K ., David, L. R., Glen W. and Jacqueline, L. (2015). Mining-caused changes to habitat structure affect amphibian and reptile population ecology more than metal pollution. *Ecological Applications*, 2240–2254.

Scandalios, J. G. (2005). Oxidative stress: molecular perception and transduction of signals triggering antioxidant gene defenses. *Brazilian Journal of Medical and Biological Research*, 995-1014.

Schaumburg, L. G., Siroski, P. A., Poletta, G. L. and Mudry, M. D. (2016). Genotoxicity Induced by Roundup® (Glyphosate) in Tegu Lizard (*Salvator merianae*) Embryos. *Pesticide Biochemistry and Physiology*, 71-78.

Siroski, P.A., Bassetti, L. A. B., Piña, C. and Larriera A. (2020). *Caiman latirostris*. The IUCN Red List of Threatened Species 2020: e.T46585A3009813. En: https://www.iucnredlist.org/api/v4/assessments/ 3009813/distribution_map/jpg. Acceso diciembre 2020.

Sujiwattanarat, P., Pongsanarakul, P., Temsiripong, Y., Temsiripong, T., Thawornkuno, C., Uno, Y., Unajak, S., Matsuda, Y., Choowongkomon, K. and Srikulnath, K. (2016). Molecular cloning and characterization of Siamese crocodile (*Crocodylus siamensis*) copper, zinc superoxide dismutase (CSI-*Sod*) gene. *Comparative Biochemistry and Physiology, Part A*, 187-195.

Tainer, J. A., Getzoff, E. D., Beem, K. M., Richardson, J. S., Richardson, D. C. (1982). Determination and analysis of the 2 Å structure of copper, zinc superoxide dismutase. *Journal of Molecular Biology*, 181–217.

Todd, B. D., Willson, J. D. and Gibbons, J. W. 2010. "The global status of reptiles and causes of their decline." In: *Ecotoxicology of Amphibians and Reptiles*, edited by Sparling, D. W., Linder, G., Bishop, C. A. and Krest, S, 47-68. Taylor and Francis, Hoboken.

Ullah, S. and Zorrieh-Zahra, M. J. (2015). Ecotoxicology: A Review of Pesticides Induced Toxicity in Fish. Advances in Animal and Veterinary Sciences, 40-57.

Valdivida, P. A., Tania Zenteno-Savín., S. C. Gardner, A. and Alonso A. (2007). Basic oxidative stress metabolites in eastern Pacific green turtles (*Chelonia mydas agassizii*). *Comparative Biochemistry and Physiology, Part C*, 111–117.

Wan, Q. H., Pan, S. K., Hu, L., Zhu, Y., Xu, P. W., Xia, J. Q., Chen, H., He, G. Y., He, J., Ni, X. W., Hou, H. L., Liao, S. G., Yang, H. Q., Chen, Y., Gao, S. K., Ge, Y. F., Cao, C. C., Li, P. F., Fang, L. M., Liao, L., Zhang, S., Wang, M. Z., Dong, W. and Fang, S. G. (2013). Genome analysis and signature discovery for diving and sensory properties of the endangered Chinese alligator. *Cell Research*, 1091–1105.

Weidinger, A. and Kozlov, A. V. (2015). Biological activities of reactive oxygen and nitrogen species: oxidative stress versus signal transduction. *Biomolecules*, 472–484.

White, G. J. 2004. "Selection of ecological indicators for monitoring terrestrial systems." In: *Environmental Monitoring*, edited by Wiersma G. B, 263-282. CRC Press, Boca Ratón, EUA.

Biographical Sketch

Odetti Lucia Magdalena

Affiliation: Cátedra de Toxicología, Farmacología y Bioquímica Legal, Facultad de Bioquímica y Ciencias Biologícas, Universidad Nacional del Litoral (FBCB-UNL), Santa Fe, Argentina. Consejo Nacional de Investigaciones Científicas y Técnicas (CONICET), CABA, Argentina.

Education:

- PhD in Biological Science, FBCB-UNL (2021).
- Bachelor of Biodiversity, Facultad de Humanidades y Ciencias, Universidad Nacional del Litoral, Argentina (2015).

Research and Professional Experience: Postdoctoral student with CONICET Scholarship at Cátedra de Toxicología, Farmacología y Bioquímica Legal, FBCB-UNL. Her doctoral thesis is titled "identification and development of gene expression markers and specific genotoxic damage to evaluate the effect induced by pesticides in *Caiman latirostris.*" Working on the evaluation of genotoxicity, gene expression patterns, and DNA global methylation as indicators of environmental stress caused by pesticides exposure in *C. latirostris*.

Professional Appointments: Posdoctoral student at CONICET and Assistant Professor at Universidad Nacional del Litoral, Santa Fe, Argentina. luodetti@gmail.com.

Publications from the Last 3 Years:

López González, E. C., Odetti, L. M., Latorre, M. A., Ávila, O. B., Contini, L. E., Siroski, P. A., Poletta G. L. 2022. A comprehensive approach using multiple biomarkers to detect damage induced by pesticides in broad-snouted caiman (*Caiman latirostris*) under *ex situ* conditions. *Heliyon*. https://doi.org/10.1016/j.heliyon.2021.e08667

López González, E. C., Odetti, L. M., Poletta, G.L., Denslow, N., Kroll K.J., Siroski, P., Parachu Marco, M. V. 2021. Optimizing protocols for high-quality RNA extraction from blood and liver tissues of *Caiman latirostris* (Broad-snouted caiman). *Russian Journal of Herpetology*. doi: 10.30906/1026-2296-2021-28-4-197-204.

Odetti, L. M., Paravani, E. Simoniello, M. F. Poletta, G. L. 2021. Identification and evaluation of antioxidant and reference genes for quantitative real-time PCR in blood of *Caiman latirostris*. *Heliyon* at. https://doi.org/10.1016/ j.heliyon.2021.e06253.

Odetti, L. M., López González, E. C., Romito, M. L., Simoniello, M. F., Poletta G. L. 2020. Genotoxicity and oxidative stress in *Caiman latirostris* hatchlings exposed to pesticide formulations and their mixtures during development through incubation material. *Ecotoxicology and Environmental Safety*. https://doi.org/10.1016/j.ecoenv.2020.110312.

Paravani, E., Odetti, L. M., Simoniello, M. F., Poletta, G. L. 2020. *Caiman latirostris extracellular superoxide dismutase 3 (SOD3) mRNA, complete cds*. https://www.ncbi.nlm.nih.gov/nuccore/MN326099.1/.

Paravani, E., Odetti, L. M., Simoniello, M. F., Poletta, G. L. 2020. Molecular analysis and bioinformatic characterization of cooper, zinc - Superoxide dismutase (Cu, Zn-SOD) gene of *Caiman latirostris*. *Molecular Biology Reports*. https://doi.org/10.1007/s11033-020-05937-y.

Burella, P. M., Odetti, L. M., Simoniello, M. F., Poletta, G. L. 2018. Oxidative damage and antioxidant defense in *Caiman latirostris* (Broad-snouted caiman) exposed *in ovo* to pesticide formulations. *Ecotoxicology and Environmental Safety*. https://doi.org/10.1016/j.ecoenv.2018.06.006.

Chapter 6

Anasarca in English Bulldogs

B. Valladares-Carranza[1,*], J. L. Zamora-Espinosa[1], M. G. Arriaga-Pérez[2], C. Ortega-Santana[1], Y. E. Felipe-Pérez[2], A. Zaragoza-Bastida[3], N. Rivero-Pérez[3], L. Delgadillo-Ruiz[4], R. Bañuelos-Valenzuela[5], P. I. Gallegos-Flores[4], V. Velázquez-Ordoñez[1] and C. Bedolla-Cedeño[6]

[1]Research Center and Advanced Studies in Animal Health. Faculty of Veterinary Medicine and Zootechnics. Autonomous Mexico State University, Toluca, Mexico
[2]Private Clinic. Toluca, Mexico
[3]Academic Area of Veterinary Medicine and Zootechnics, Autonomous University of the State of Hidalgo, Hidalgo, Mexico
[4]Academic Unit of Biological Sciences
[5]Academic Unit of Veterinary Medicine and Zootechnics, Autonomous University of Zacatecas, Zacatecas, Mexico
[6]Faculty of Veterinary Medicine and Zootechnics, San Nicolas de Hidalgo Michoacan University, Morelia, Mich, Mexico

Abstract

Anasarca is a pathology in English Bulldog dogs, which is a fetal malformation in puppies that presents generalized edema. The greater accumulation of fluid occurs at the subcutaneous level and causes the loss of the normal morphology of the canine; produces dystocic deliveries, in some cases the pups are born alive but their viability is unlikely. The objective of the present work was to analyze and assess by pathological and histological studies the corpses of 3 English Bulldog puppies that were referred to CIESA for diagnosis, a necropsy of the specimens was performed, and samples were collected for histopathological study of most of the tissues, which were preserved in 10% buffered formaldehyde,

* Corresponding Author's Email: bvalladaresc@uaemex.mx.

and prepared with the H&E stain. Macro and microscopic photographic evidence of the lesions was gathered. In the anatomopathological study. In general, the alterations observed in the 3 specimens were: an increase in size with a cylindrical body and soft at palpation, when making the primary cut, ascites was found as the main finding, when approached with a cut. Secondary, hydrothorax, pulmonary hypoplasia in the thoracic cavity was found. While ascites in the abdominal cavity, pulmonary hypoplasia, generalized and hepatic congestion, congestion and intestinal mucoid content was also observed. The histopathological study, in the lung revealed: alveolar collapse, loss of the structure of the blood vessels, the wall of the blood vessels presented different degrees of injury from those that present hydropic degeneration to necrosis of smooth muscle cells, endothelial cell hyperplasia; however, the bronchi did not show defined morphology and structures compatible with hyaline cartilage were not identified, as well as peribronchial and peribronchiolar lymphoid aggregates were not identified. The heart presented undefined myocytes, a scarce amount of trabeculae was observed, also the direction of the cardiac fibers was found in different directions, differing from the normal histological pattern. The results found in the microscopic study indicate that the tissues of the different organs are not fully mature, thus denoting their incompatibility with life. This pathology is very frequently associated with cleft palate. The diagnostic method to be used is ultrasound, a tool that would more accurately indicate the excess fluid in the fetus.

Keywords: anasarca, English Bulldog, ascites, dog

Introduction

Anasarca in dogs, also known as the walrus puppy syndrome or "walrus," is a process of extravascular fluid collection in the neonate that occurs in the gestational period. Anasarca is one of the most common causes of fetal dystocia; it's treatment consists of increasing oncotic pressure, supplementing proteins in the form of frozen colostrum and plasma. This entity appears mainly in the English Bulldog, although it´s etiology has not been fully established, three possible causes are considered: Cardiac origin, which occurs due to an increase in intravascular pressure, with a decrease in oncotic pressure. Protein origin, due to protein malabsorption, malabsorption or neonatal liver failure. Embryonic and hereditary origin, due to recessive genes. Congenital defects are those that are present at birth without being able to specify the cause. In some circumstances they are genetic in origin; sometimes teratogenic alterations related to maternal malnutrition may be responsible; it has been reported that this type of puppy survives from one to three days later, puppies could normalize it´s condition normalize and survive, but in other cases it is reported that the clinical picture is complicated, causing the death of the puppy within a week of life (Noakes et al., 2009). Toxic origin: Due to the administration of estrogenic hormones or the use of anticonvulsants before gestation can cause an inadequate formation of the lymphatic system in puppies. In turn, it has also been reported to occur with transplacental infection by Parvovirus type 1 (ECODVet, 2010).

Edema

It is considered the process of abnormal accumulation of serous fluid in various organs, cavities or tissues of the body (Dog Breed Info Center, 2011). This increase in the extravascular compartment of the extracellular fluid can occur in a localized or generalized distribution. It is important to consider that this liquid contains water, electrolytes and very few proteins and is called transudate; it should not be confused with exudate, which is also an interstitial fluid or in cavities that contain greater amounts of proteins and are usually of inflammatory origin (Lorenz et al., 2009).

In the macroscopic aspect of this process, it can be seen that the tissue increases in thickness and, when cut, the transudate flows, which is transparent or yellowish and generally does not coagulate. When found in cavities it can cause, depending on its location, heart, lung or digestive disorders. When there is edema in the lung, this organ does not collapse when opening the thoracic cavity and a foamy fluid is often found in the bronchi and trachea. In subcutaneous tissue, to verify that it is edema, the clinician uses the godet sign.

Regarding its microscopic appearance: In a histological section stained with hematoxylin-eosin (H&E), the liquid appears as a translucent or pink-colored material that extends in the interstitial connective tissue, in perivascular spaces. When the edema is pulmonary, the alveoli are distended and filled with this eosinophilic material (Jones and Duncan, 1990).

The cause of edema is considered to be a disorder in the exchange of fluids and other molecules between the blood capillaries and the extravascular (interstitial) tissue. The mechanism of pinocytosis has also been proposed to explain fluid leakage, although it appears to play a minor role in edema formation. The four main factors that determine whether fluid will flow out of the capillary into the interstitial space or in the opposite direction are:

1. Intravascular hydrostatic pressure (HP); 2. Hydrostatic pressure of interstitial fluid (HPI); 3. Plasma colloid osmotic pressure (PCOP), and 4. Interstitial fluid colloid osmotic pressure (ICOP)

The balance of fluid exchange between vascular and interstitial compartments is essentially maintained by means of physicochemical pressure principles: hydrostatic and colloid osmotic, where the capillary wall acts as a semi-permeable membrane. Hydrostatic pressure is the mechanical force given by the pumping of the heart; while the colloid osmotic pressure is the attraction exerted on the water by the solutes dissolved in it (Mc Gavin and Zachary, 2007).

Capillary pressure is difficult to measure and is not known in most animal species. At the capillary arterial level, a hydrostatic pressure of 30 mmHg and a colloid osmotic pressure of 25 mmHg are accepted; the latter is made up of protein molecules that do not cross the capillary wall.

It must be taken into account that there is also a normal hydrostatic pressure in the interstitial fluid (HPIF), which is approximately 8 mmHg, and another colloid osmotic pressure (COP), of 10 mmHg in normal tissue; on the one hand, HP (30 mmHg) pushes water out of the capillary and HPIF (8 mmHg) pushes it in, so the net balance is that a little water comes out (30-8 = 22 mmHg).

On the other hand, the CP pulls water into the capillary (25 mmHg) and the CPI towards the interstitial tissue (10 mmHg) and the balance of the colloid osmotic pressures, attract a little water towards the capillary (25-10 = 15 mmHg). Most of the water that leaves the capillary, due to the difference in hydrostatic pressure, returns to the capillary, due to the difference in colloid osmotic pressure. The excess water is drained by the lymphatic system. When hydrostatic or colloid osmotic pressures, or both, are altered, the balance is disturbed and the consequence will be edema (Dunlop and Malbert, 2004).

Thus, the increase in hydrostatic pressure will produce a greater filtration of the liquid into the interstitial space. This liquid will not be able to be reabsorbed in the same amount by the venous and lymphatic capillaries, which is why it will form edema.

An increased hydrostatic pressure is generally reported in heart disease. In the event of left ventricular failure, an edema is initially formed in the lungs and when the failure is located in the right ventricle, generalized systemic edema will appear, due to the general staging of venous blood. However, these defects on one side or the other of the heart will soon have repercussions throughout the entire body (Patruck et al., 1998).

During the early stages of congestive heart failure and the resulting decrease in glomerular filtration, there will be an increase of renin secretion and subsequent angiotensin activation, as well as sodium retention. Due to the congestive disorder, which also involves the liver, it does not catabolize aldosterone, whose concentration increases in the circulating blood. These mechanisms, among others, initially compensate for cardiac dysfunction, but if the condition persists, the compensatory mechanisms are not sufficient, renal failure worsens and with it sodium retention, which causes an increase in hydrostatic pressure (Mc Gavin and Zachary, 2007).

Decreased colloid-osmotic (oncotic) pressure. The edema caused by the low colloid-osmotic pressure (oncotic), has its origin in hypoproteinemia. The concentration of plasma proteins, especially albumin, decreases due to malnutrition or poor absorption and loss of plasma proteins or failure in their synthesis.

The leading causes of malnutrition or malabsorption, include the lack of protein in the diet, intestinal parasites, and chronic gastrointestinal diseases that lead to chronic diarrhea (Johne's disease).

Also the important causes of plasma protein losses are: Chronic kidney diseases that present with albuminuria, and chronic liver diseases, due to the inability of the liver to synthesize proteins (Trigo et al., 2004).

Obstruction of lymphatic circulation; caused by processes, such as: tumors, abscesses, granulomas, paratuberculosis in sheep mesentery, *Wuchereria bancrofti* in human elephantiasis, among others (Jones and Duncan, 1990).

Sodium and water retention (renal epithelium). It occurs in kidney diseases when tubular reabsorption does not take place (interstitial nephritis), or in cases of pulmonary disorders (excess antidiuretic hormone, mineralocorticoids).

Increased capillary permeability (anorexia, shock, anaphylaxis, trauma, inflammation). Increased capillary permeability can have several causes, including hypoxic endothelial damage in chronic congestion, by bacterial toxins (*Escherichia coli* enterotoxemia, clostridial toxemia, *Pasteurella spp* septicemia and other chemical toxins, for example, ANTU = alpha naphthyl thiourea) and allergic processes; in most of these examples, the liquid contains significant amounts of proteins that come out of the capillary and, additionally, hypoxia occurs, in the case of chronic congestion, or toxic factors. Therefore, in many cases, the fluid no longer corresponds to a transudate and is called serous or fibrinous exudate, which is almost always the product of an inflammatory process (Trigo et al., 2011).

Table 1. Classification of edematous fluids, according to their location

Type	**Affected tissue or site**
Anasarca	Tejido subcutáneo (edema subcutáneo generalizado)
Hydroperitoneum (ascites)	Peritoneal cavity
Hydrothorax	Thoracic cavity
Hydropericardium	Pericardial sac
Hydrocele	Scrotum
Hydrocephalus	Cranial cavity (sistema ventricular system, meningeal spaces or both)

Modified from Trigo et al., 2004.

Peripheral Edema

This process is considered due to the abnormal accumulation of interstitial fluid in the subcutaneous tissues, which is why the following types can be distinguished:

1. Located (in one limb).
2. Regional (anterior or posterior limb).
3. Generalized (usually in the midline, on sloping or dependent surfaces).

Physiopathological, it occurs due to a reduced plasma oncotic pressure (hypoproteinemia, albumin content less than 1.5 g/dL), high vascular hydrostatic pressure, high capillary permeability and abnormality of lymphatic drainage; for example; dysgenesis, obstruction and destruction (Patruck et al., 1998).

The affected systems are: integument / exocrine - in which swelling of the subcutaneous tissues occurs, especially in ventral or sloping areas: serum exudation or dermal ulceration in the affected limbs, skeletal muscle - ambulatory abnormalities if the swelling of the limb is excessive. With a history of recent trauma (including surgery), exposure to ticks, poisonous insects, arachnids, snakes, or a history of heart or vascular disease. With data from the physical examination, localized or diffuse subcutaneous swelling is observed including extremities and ventral midline, in the affected limbs there is usually no pain, but if the swelling is pronounced there may be pain; serum exudation from swollen limbs, dermal ulceration and evidence of autotrauma of affected areas (can lead to necrosis), subcutaneous edema may or may not be soft (depressible), local or regional lymph node enlargement, exudative tracts (animal with foreign body or my ketogenic infections) (McGavin and Zachary, 2007).

Among the causes or origin of this type of edema (a single limb) are: high hydrostatic pressure, acute arterial or venous thrombosis, arteriovenous fistula, high capillary permeability, angioedema due to the bite or sting of snakes, insects or arachnids and a reaction in the injection site of a drug, trauma, burns, neoplastic infiltration, cellulitis (e.g., foreign body, infection and inflammation), lymphatic abnormalities, local or subcutaneous lymph node neoplastic infiltration (due to local or metastatic tumor), traumatic lymphatic disruption, lymphangitis and congenital malformation of the lymphatic system (Dunlop and Malbert, 2004).

In the case of regional peripheral edema (thoracic limbs and pelvic limbs or generalized: reduced oncotic pressure (hypoalbuminemia), liver disease, protein-losing gastrointestinal or renal disease (nephrotic syndrome), substantial blood loss with replacement of non-colloid fluids, high blood pressure hydrostatic, congestive heart failure (CHF), biventricular or right, pericardial effusion with tamponade, right atrial mass, cava vein obstruction, overhydration especially if associated with oliguric acute renal failure or hypoproteinemia, high capillary permeability, sepsis, vasculitis (in infectious disease and tick-related pathology), and some snake venoms, trauma, burns, lymphatic abnormality (lymphedema), infiltration or abscessation of regional or multiple lymph nodes, traumatic lymphatic disruption, lymphangitis, and congenital malformation of the lymphatic system (Jones and Duncan, 1990; Mc Gavin and Zachary, 2007).

In facial edema: thrombosis with jugular catheter, tight cervical bandage or brace, myxedema (related to hypothyroid states), local lymphadenopathy, juvenile pyoderma (Patruck et al., 1998).

Differential diagnosis should be made with peripheral edema associated with myxedema and inflammation (cellulitis), which is usually not soft. If the animal has peripheral edema with jugular venous distention and ascites, cardiac or pericardial disease should be considered. If the animal has edema in both thoracic limbs, jugular distention, and ascites, consider cranial cava vein obstruction. If there is edema of both lower extremities, ascites, and no jugular distention, consider hypoproteinemia or obstruction of the caudal vena cava. In bilateral lower extremity edema without ascites, consider lymphatic or vascular obstruction. Hyperglobulinemia can be associated with infectious causes. Edematous swelling with palpable thrill, consider arteriovenous fistula. Stingers from snake bites may be visible. Petechiae or ecchymoses, consider disseminated intravascular coagulation (DIC), thrombocytopenia, and vasculitis. Arteriovenous fistulization and lymphatic obstruction can develop at sites of previous trauma (Patruck et al., 1988; Dunlop and Malbert, 2004).

Anasarca in English Bulldog Dogs (Water Puppy Syndrome)

The anasarca problem, which implies a diffuse edematous state of the subcutaneous tissue, is a condition whose pathogenesis is not completely defined; and in dogs, the highest incidence is reported in English Bulldogs compared to other canine breeds. A similar condition has been described in humans. In dogs it has been associated with Canine Minute Virus (CMV), which is a type of virus of the Parvoviridae family that affects canines in which histopathological lesions suggestive of myocarditis and hyperplasia in the spleen and thymus have been reported. In sheep and cattle, the pathogenesis seems to be related to an autosomal recessive gene, while in other animals, this has not been verified, although the high predisposition in some dog breeds indicates a genetic defect. The problem of generalized edema has been associated with multiple causes, which can refer to right heart failure, pericardial effusion, hypoproteinemia, altered vascular permeability, and lymphatic obstruction, mainly. The anasarca problem has been theoretically related to cardiac anomalies, with a traumatic origin such as placental disruption, and with viral infection or malformations of the lymphatic drainage system (Castro et al., 2011).

Affected dogs are usually newborn puppies (they are always achieved by cesarean section due to their size), because they are twice as large as usual, and because of the enormous fluid retention they present under the skin (anasarca). On many occasions these puppies also present other abnormalities such as a cleft palate. In affected females, mild edema is sometimes seen before delivery in the mammary tissue, around the nipple. Water pups are usually born alive, but in the most severe cases they die within a few hours of delivery due to the lack of oxygen supply through the umbilical cord. Either the entire litter can be affected or just a few pups. Mild cases that survive 2-3 days after birth, usually recover completely (Ferrer et al., 2008; Peterson and Kutcher, 2011).

Anasarca in English Bulldogs can affect a single individual, producing generalized edema and fluid accumulation (Johnston et al., 2001). In the case of diffuse edema in the affected fetus, it increases not only its size, but also causes the fetus to have a soft consistency that hinders delivery. On rare occasions, natural delivery of one of these animals is achieved (Jones and Joshua, 1984).

Fetal Anasarca

The affected fetus regularly reaches term and concern is increased at delivery, due to the large increase in fetal volume caused by excess fluid in the subcutaneous tissue, particularly of the head and extremities; in the case of the head, there is not so much inflammation that the normal characteristics are masked and the resulting appearance is quite grotesque, the enormous swelling of the extremities that they present is very visible. There is often excess fluid in the peritoneal and pleural cavity with dilatation of the umbilical cord and inguinal rings, as well as hydrocele (Noakes et al., 2009).

Susceptible Breeds

There are various opinions about the development of this process in puppies, some consider that it is congenital or due to the effect of the environment, diet, trauma of the pregnant female, or even a virus. There is no evidence to show that it is hereditary, but it does affect some breeds more than others, especially the English Bulldog and other brachycephalic breeds including: Boston Terriers, Shar Pei, Bullmastiff, Basset Hound, Chow Chow, Wire Fox Terrier, West Highland White Terriers, Cavalier King Charles Spaniels and Pugs (Burke, 2001; Crusheds bulldogs, 2007; Callisto, 2009). It is less frequently found in dogs of breeds such as; Borzoi, Maltese, Labrador Retriever, Schnauzer, Bichon Frise, Chihuahuas, German Shepherd, Old English Sheepdog (Bobtail), Poodle, Great Dane, German, Pointer, mixed dogs and in a zoo wolf (Dog Breed Info Center, 2011).

English Bulldog

The Bulldog is thought to descend from the Molossians of Epirus, war dogs introduced to England by Phoenician seafarers. As its name indicates, it was raised to fight against bulls. Fights between dogs were also organized. In 1835, this cruel practice was abolished. In 1875, the first standard was published. Subsequently, the elections that took place made the Bulldog a companion dog. Description of the breed. Head: Solid, short face, wide skull, loose and wrinkled skin, short wide and retracted snout, wide nostrils, thick and hanging lips, wide and square jaws, the lower jaw exceeds the upper and curves upwards. Eyes: wide apart, of medium size, round and very dark. Ears: wide apart, set high on skull; small, fine, rose-shaped (folded back to reveal the inside of the pinna and external ear canal). Body: short, very compact. Very wide neck with dewlap, broad and slanting shoulders, large rounded chest, short strong back, raised loin, sprung ribs, raised strong hindquarters, raised belly. Strong and very muscular limbs, widely separated forelimbs, round and compact feet, the forelimbs are very slightly turned outwards.

Tail: set low, round, of moderate length, hanging, without marked curve upwards. Hair: fine, short, smooth and tight. Coat: monochrome or "SUIE," (that is, monochrome with a black mask or muzzle). Solid colors: red, fawn, brindle, white or mixed (white with the preceding colors). Liver, black and tan colors are not accepted. Height: from 30 to 40 cm. Weight: male from 24 to 25 kg and female from 22 to 23 kg. Character, aptitudes, education: lively, intrepid, brave of character, balanced and trustworthy. Despite his fearsome appearance, he is affectionate in nature, quiet, good-natured, and very quiet. It is an excellent companion for children. It must be educated firmly (Gallitelli et al., 2003).

Method

An exhaustive search for information was carried out in different sources in which the characteristics of the disease were analyzed and described in detail, writing and detailing the process with the material obtained from 3 cases referred to the diagnostic area of the Center for Research and Advanced Studies. in Animal Health (CIESA), of the Faculty of Veterinary Medicine and Zootechnics of the UAEM. In the area of pathology and histopathology.

Results

For the present study, 3 English Bulldog puppies (Figure 1) were sent to the diagnostic area of the Center for Research and Advanced Studies in Animal

Health (CIESA) of the Faculty of Veterinary Medicine and Zootechnics of the UAEM. To which the necropsy was performed, with the procedure described by Aluja and Constantino (2002), under biosafety conditions using the necessary material and equipment.

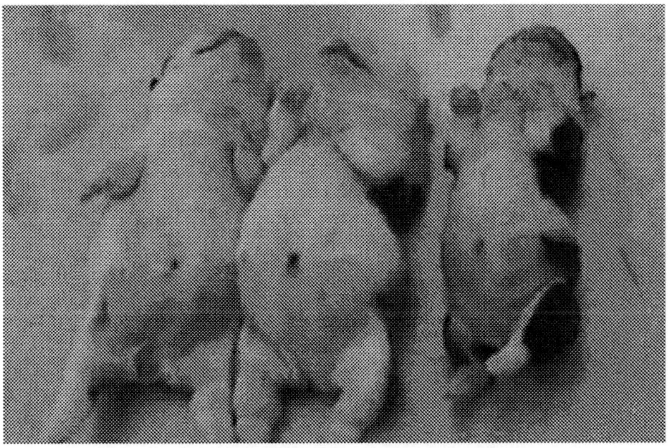

Source: M en C José Luis Zamora E./ MVZ. Guadalupe Arriaga P.

Figure 1. English Bulldog puppies with abdominal distention.

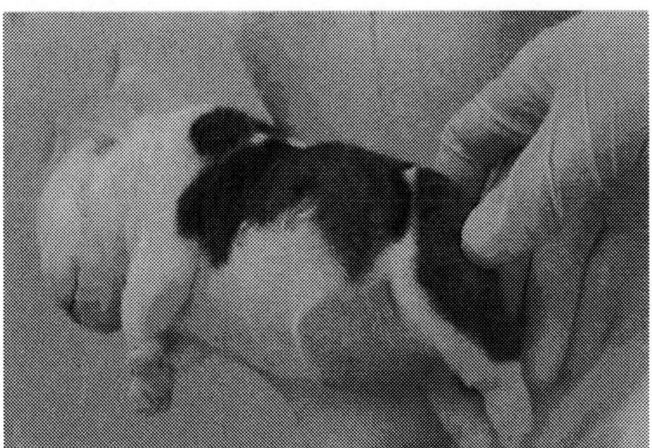

Source: M en C José Luis Zamora E./ MVZ. Guadalupe Arriaga P.

Figure 2. English Bulldog puppy with soft consistency - generalized subcutaneous edema.

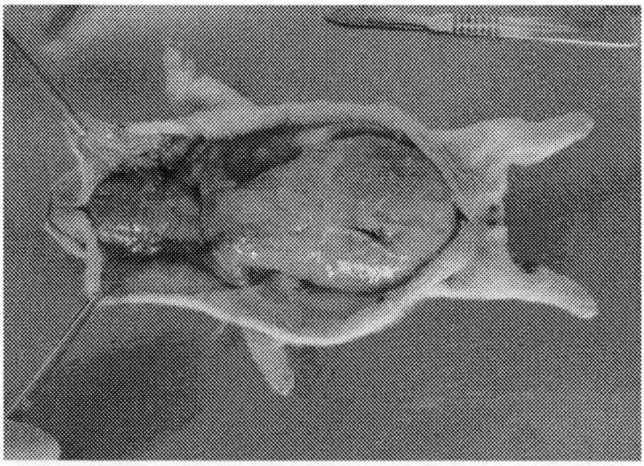

Source: M en C José Luis Zamora E./ MVZ. Guadalupe Arriaga P.

Figure 3. English Bulldog puppy. Generalized edema of subcutaneous tissue.

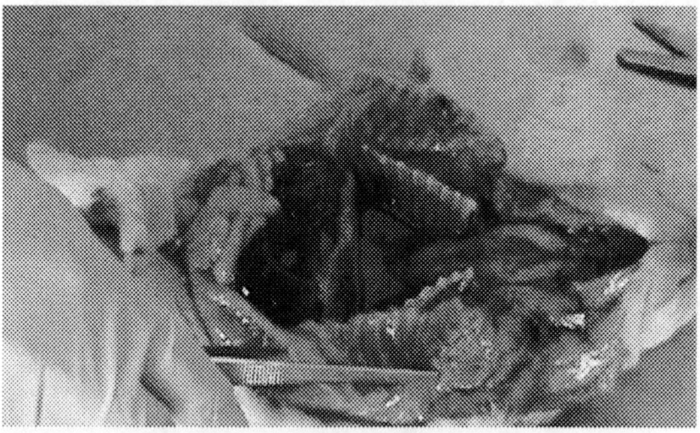

Source: M en C José Luis Zamora E./ MVZ. Guadalupe Arriaga P.

Figure 4. English Bulldog puppy with hydrothorax and ascites.

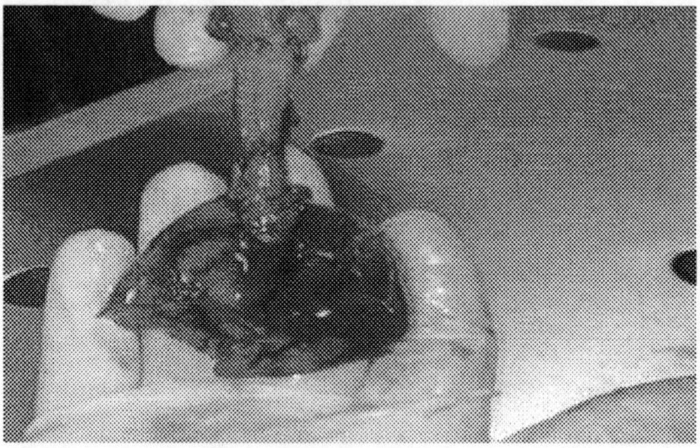

Source: M en C José Luis Zamora E./ MVZ. Guadalupe Arriaga P.

Figure 5. English Bulldog puppy lungs, with marked hypoplasia, congestion, and edema.

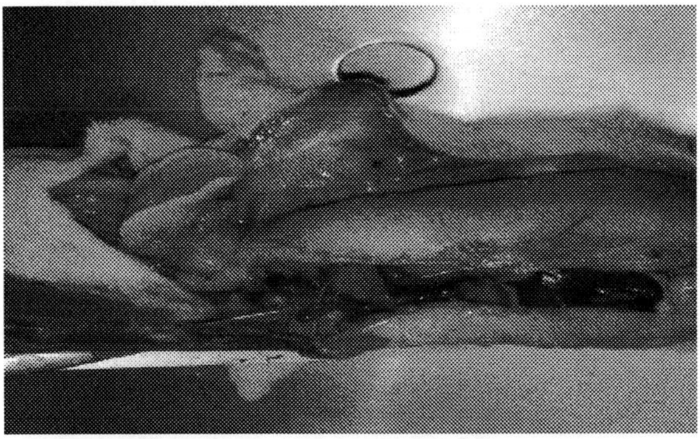

Source: M en C José Luis Zamora E./ MVZ. Guadalupe Arriaga P.

Figure 6. English Bulldog puppy corpse. With generalized congestion of subcutaneous tissue and edema.

Source: M en C José Luis Zamora E./ MVZ. Guadalupe Arriaga P.

Figure 7. Liver congestion in an English Bulldog puppy.

In the anatomopathological study, the alterations observed were: at the macroscopic inspection of the 3 specimens, an increase in size with a cylindrical body and soft on palpation (Figure 2), when making the primary cut, ascites was found as the main finding (Figure 3), when approaching the thoracic cavity with a secondary cut, the following was found: accumulation of fluid in the thoracic cavity (hydrothorax) in the three puppies, pulmonary hypoplasia, ascites in the abdominal cavity (Figure 4), pulmonary hypoplasia (Figure 5), generalized congestion (Figure 6), and hepatic (Figure 7), congestion and intestinal mucoid content (Figure 8).

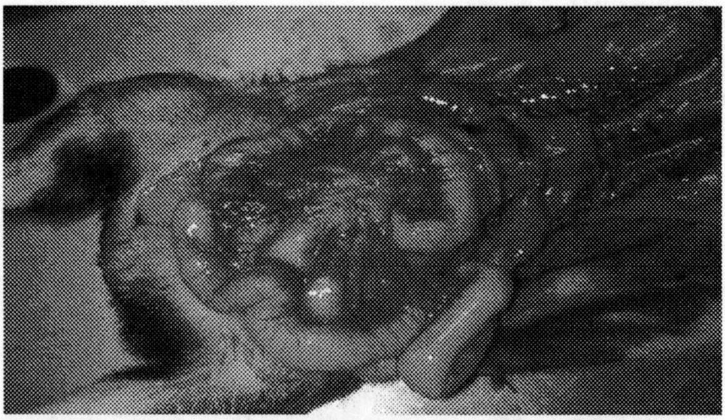

Source: M en C José Luis Zamora E./ MVZ. Guadalupe Arriaga P.

Figure 8. Congestion and presence of intestinal mucoid content.

Parallel to the anatomopathological study, samples were collected for histological study, and the results of the analysis of the preparations with Hematoxylin & Eosin staining were as follows:

The microscopic findings: in the lung, the following were observed: alveolar collapse, loss of the structure of the blood vessels, the wall of the blood vessels presented different degrees of injury from those that presented hydropic degeneration to necrosis of smooth muscle cells, hyperplasia of endothelial cells; the bronchi did not show a defined morphology and no structures compatible with hyaline cartilage were identified, which gives the impression that there was no formation of them, as well as no peribronchial and peribronchiolar lymphoid aggregates.

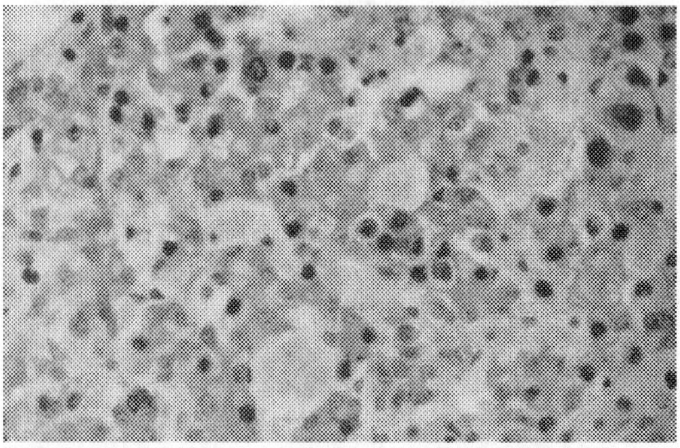

Source: M en C José Luis Zamora E./ MVZ. Guadalupe Arriaga P.

Figure 9. Liver of an English Bulldog puppy. Histopathological study. Presence of hydropic degeneration of hepatocytes (H&E stain).

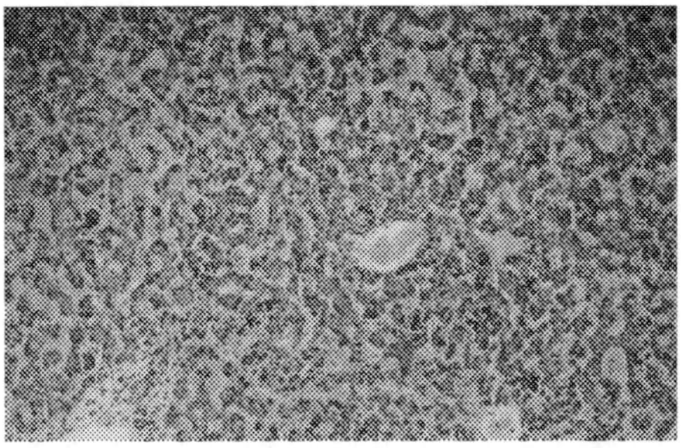

Source: M en C José Luis Zamora E./ MVZ. Guadalupe Arriaga P.

Figure 10. Liver. Loss of the structure of the hepatic cords and diffuse congestion is identified (H&E stain).

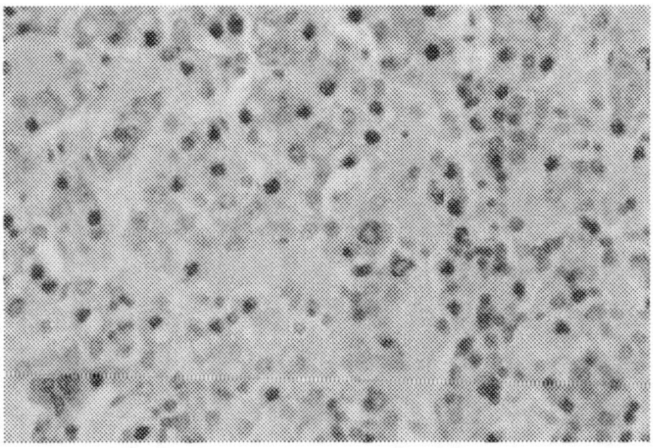

Source: M en C José Luis Zamora E./ MVZ. Guadalupe Arriaga P.

Figure 11. Liver. Morphology and undefined structure of hepatocytes; without hepatic cords, hydropic degeneration and absence of central vein (H&E stain).

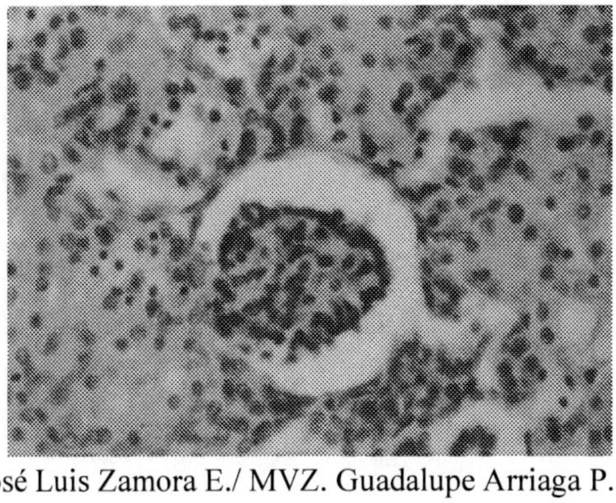

Source: M en C José Luis Zamora E./ MVZ. Guadalupe Arriaga P.

Figure 12. Kidney. Incomplete formation of the glomeruli (H&E stain).

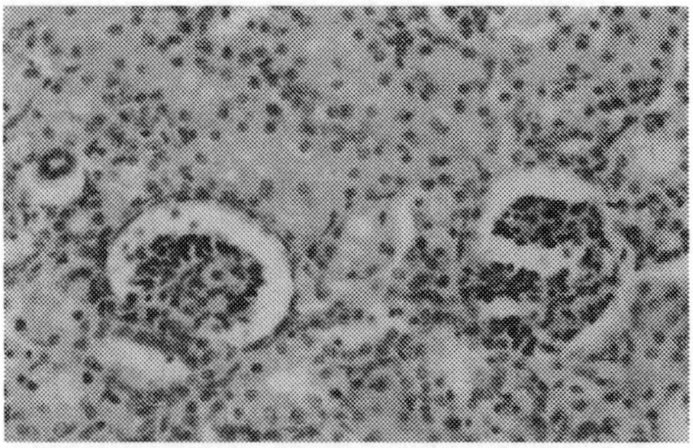

Source: M en C José Luis Zamora E./ MVZ. Guadalupe Arriaga P.

Figure 13. Kidney. It is observed that the macula densa of the glomerulus is very prominent and poorly structured (H&E stain).

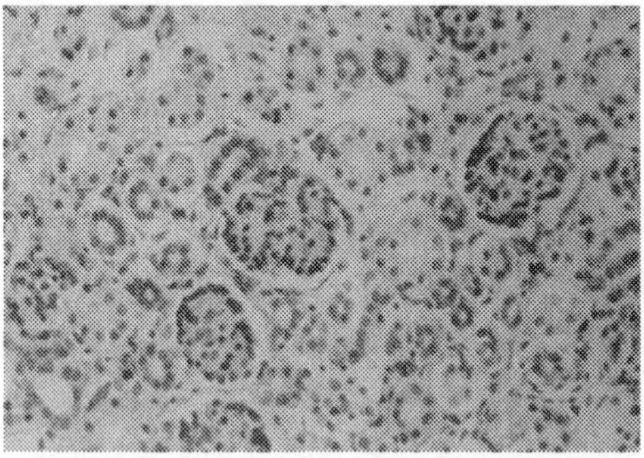

Source: M en C José Luis Zamora E./ MVZ. Guadalupe Arriaga P.

Figure 14. Kidney. Glomeruli without macula densa, amorphous urinary spaces (H&E stain).

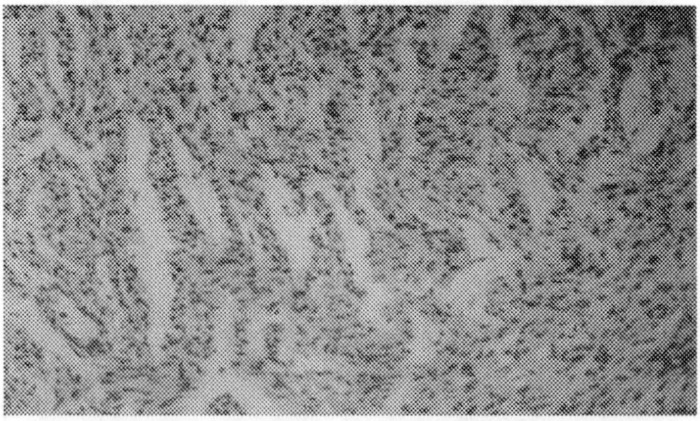

Source: M en C José Luis Zamora E./ MVZ. Guadalupe Arriaga P.

Figure 15. English Bulldog puppy heart. Variation of the direction of cardiac fibers (H&E staining).

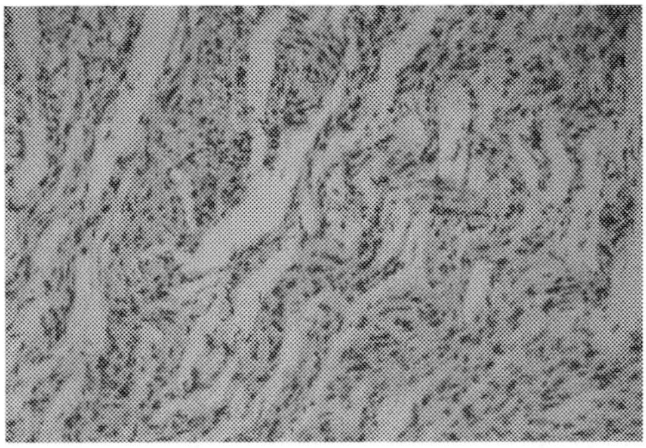

Source: M en C José Luis Zamora E./ MVZ. Guadalupe Arriaga P.

Figure 16. English Bulldog puppy heart. Nuclear pleomorphism of myocytes (H&E stain).

Source: M en C José Luis Zamora E./ MVZ. Guadalupe Arriaga P.

Figure 17. Ganglion. Absence of lymphoid follicles (H&E stain).

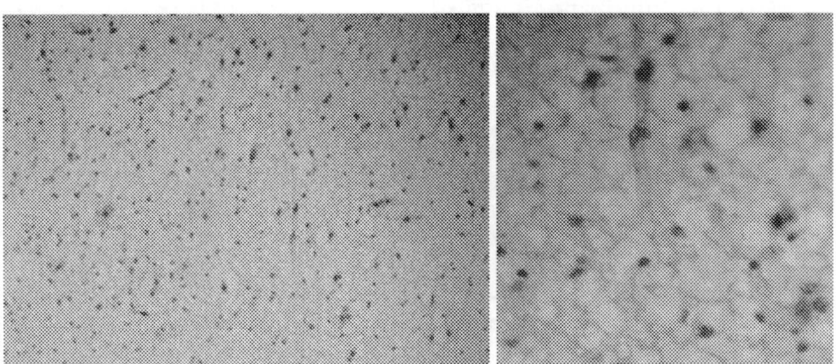

Source: M en C José Luis Zamora E./ MVZ. Guadalupe Arriaga P.

Figures 18 and 19. Brain. Neuropil vacuolization.

In the heart, undefined myocytes were observed, with a small amount of trabeculae and very thin in cardiac muscle, the direction of the cardiac fibers was in different directions, discerning the normal histological pattern (Figure 15 and 16).

The histopathological study of the lymph node revealed the presence of thin trabeculae and loss of the normal histological structure (Figure 17).

Likewise, in the nervous tissue of the three pups analyzed, a generalized vacuolization of the tissue was observed (Figures 18 and 19).

Discussion

In the present study in which 3 English Bulldog puppies were sent for diagnosis to the CIESA-UAEM, due to the clinical history, as well as the general macroscopic observation of the corpses, an alteration in the development of the specimens was considered, which in the anatomopathological study, due to the observed alterations, the anasarca problem was determined as the determining pathology that caused the death of these animals.

Fetal anasarca (*hydrops fetalis*) corresponds to generalized edema in the subcutaneous tissue, accompanied by accumulation of fluid in the thoracic and abdominal cavities, which causes dystocic deliveries due to disproportion of the fetus (Osorio et al., 2011).

The causes that have been described for congenital fetal anasarca are prenatal cardiovascular anomalies of genetic origin, trauma with placental abruption, and canine parvovirus infections in the last third of gestation (Osorio et al., 2011; Celticpride, 2011).

The macroscopic and microscopic evidence of anasarca, particularly in the English Bulldog species, makes it necessary to consider the feasibility to methodologically evaluate the cause or origin of the process.

It has become evident that the canine breed evaluated in the study is one in which it can occur more frequently, and in which the reporting of cases may increase soon, however the care of the patient in which it has been diagnosed at some time must take the necessary precautionary measures to avoid its new presentation through management, all under strict supervision of the Veterinary Physician.

As a fetal pathology, it is a cause of dystocia and its presentation is related to a congenital malformation of the lymphatic system, especially due to the administration of certain drugs (hormones, anticonvulsants), which can be used to treat or prevent maternal diseases that carry a probable or known risk for use during gestation. The studies carried out by Minovich et al., (2004), report that fetal anasarca has a close relationship with teratogenesis, fetal death and abortions in felines.

The information that Sucupira et al., (2011), have collected from cases of anasarca in sheep, refer that the pathophysiology of anasarca is not yet fully determined and mentioned that the characteristic lesions of congenital anasarca are generalized subcutaneous edema and the transudate in the thoracic or abdominal cavity, macroscopically speaking. The cases of fetal anasarca, also called *hydrops foetalis*, are not well recognized in sheep, the main sign observed lies in the difficulty of the ewes to give birth; lambs born are up to four times normal size due to widespread subcutaneous edema; in affected animals surviving parturition, painless diffuse swelling (fetal anasarca) may be seen.

While in cattle there is information obtained by García et al., (2012), in their studies have documented malformations characterized by *hydrops fetalis* such as hydrocephalus, fetal ascites and fetal anasarca, these being incompatible with life, although we realize that they are very similar to the fetal anasarca of English Bulldog puppies.

According to the findings in the present study, the observed alterations match with the poor development of the puppies, considering that once the poor definition of structures in both the anatomy as well as in the physiology are interrelated to the appearance of the problem.

Data referred to animal species, in which the pathology of the process has been documented in a certain way, states that in a general way is a circulatory failure, although it is not fully elucidated, it may be due to a variety of factors, in the present case it cannot be attributed to one in particular, but possibly it is due to the breed of the specimens evaluated; in which genetic factors can contribute to the appearance of anasarca.

In humans, *hydrops fetalis*, is a serious problem that puts the life of the baby at risk, it is characterized by causing generalized edema; depending on its origin this problem has two types; the immune hydrops, which can develop when there is an incompatibility of the Rh (negative) system of the mother, and the Rh (positive) of the fetus causing a reaction of the immune system, that is, the mother's antibodies, which cause the destruction of red blood cells of the fetus causing anemia, and as the baby's organs are not able to compensate for the anemia, hydrops can develop, manifesting itself in the fact that the heart begins to develop failure and large amounts of fluid accumulate in the fetal tissues and organs. The following can be found among the diseases or complications associated with non-immune dropsy: severe anemia, congenital infections, lung or heart defects, chromosomal abnormalities and/or congenital defects and liver diseases (Infogen, 2013).

Presentation of pathologies such as the one addressed in this case, makes us consider that diagnostic techniques and methodologies must be increasingly effective to detect their cause or origin in a timely manner; currently new agents may play a relevant role, which will produce deaths and economic losses to the affected species; however, in an effort to preserve, prevent and control this kind of alterations in Veterinary Medicine, even timely care services in animal health must be improved.

As a pathology, anasarca has been reported by very few doctors, on many occasions it has gone underdiagnosed or with a diagnosis of lymphedema (Patruck et al., 1998), in order not to fall into this detail it is necessary to make an early diagnosis of the pregnant female by means of an ultrasound, this being the most accurate and appropriate means to see the status of the fetuses and to schedule the delivery for primary care of the pregnant female. In case of lymphedema, this condition will be corrected according to the level of the process and with its proper treatment (Osorio et al., 2011).

Conclusion

Fetal anasarca is a multifactorial pathology that can affect various animal species, mainly dogs of the English Bulldog breed. Anatomopathological evaluation shows: swollen newborns, enlarged in size, cylindrical in shape and soft to the touch.

In the process there is accumulation of fluids in cavities, pulmonary congestion with or without pulmonary hypoplasia, atrial cardiac congestion, hepatic and intestinal loops congestion.

Congenital anasarca must be differentiated from generalized lymphedema.

Histopathological studies reveal a loss of the histological normal structure of most tissues (heart, brain, kidney, liver), corresponding to cells of each organ which are incomplete, amorphous and in different directions.

Ultrasound in pregnant females is the most appropriate diagnostic tool for detection of congenital fetal anasarca.

References

Aluja, A. S y Constantino, F. (2002). *Necropsy techniques in domestic animals*. 2th ed. Manual moderno, México, D. F. pp. 22-35.

Burke, M. (2001). *Water puppies*. http://www.bullwrinkle.com/Assets/ Health%20Topics/Water%20Puppies.htm. (March 24, 2021).

Calisto, C. (2009). *Fetal anasarca or congenital anasarca*. http://clinicaveterinaria diagnopet.blogspot.mx/2009/11/caso-clinico-noviembre-2009.html. (June 15, 2021).

Castro, M. J, Zamora, E. J. L., Valladares, C. B., Díaz González, B. A. E., Fajardo, M. R. (2011). Anasarca in English Bulldog puppies. *XX National Congress of Veterinary Pathologists*. Puerto Vallarta, Jalisco. México. pp. 18-21.

Celticpride (2011). *Water puppies*. http://celticpride-bulldogs.co.uk/Celticpride %20Puppy%20health.htm. (June 7, 2021).

Crusheds bulldogs. (2007). *Anasarca*. http://www.crusheds.com/veterinaria.htm# ANASARCACRUSHEDSBULLDOGS, (March 2, 2021).

Dog Breed Info Center (2011). *Water (walrus) puppies*. http://www.dogbreedinfo. com/whelpingwaterpuppies.htm. (March 2, 2021).

Dunlop, H. R., Malbert, H. C. (2004). *Veterinary pathophysiology*. Blackwell publishing, USA.

ECODVet (2010). *Anasarca in English Bulldog fetuses*. http://ecodvet. blogspot.com/2010/11/anasarca-en-fetos-de-bulldog-ingles.html (March 16, 2021).

Ferrer, E., Risco, M., Pérez, R. (2008). *Water puppies*. http://www.mibulldogingles. com/enfermedad/cachorro_agua.php (March 15, 2021).

Gallitelli, B, Legross, C., Alvez, K, Bally, R., Bloomberg, M., Bourdin, M., Bullard, B., Chatelain, Ë., Correau, J. et al. (2003). *Royal canin encyclopedia of the dog*. Aniwa Publishing, Francia.

García, J. A., Cruz, G. E., (2012). *Management, Delivery and Dystocia, Bovine Clinic*. http://mvz.unipaz.edu.co/textos/lecturas/clinica-de-los-bovinos-1/unidad-5_-enfermedades-del-aparato-reproducor/manejo_parto_y_distocia.pdf. (May 19, 2021).

Infogen. (2013). *Hydrops fetalis*. http://www.infogen.org.mx/Infogen. (February 15, 2021).

Johnston, S. D., Kustril, M. V. R., Olson, P. N.S. (2001). *Canine and feline Theriogenologia*. Saunders, Philadelphia.

Jones, C. T., Duncan, H. R. (1990). *Veterinary pathology*. 5th ed. Hemisferio sur, España.

Jones, E., Joshua, D. (1984). *Clinical canine reproductive problems*. Manual Moderno S. A. de C. V, México D. F.

Lorenz, D. M., Neer, M. T., DeMars, L. P. (2009). *Small animal, medical diagnosis*, 3th ed., Wiley-Blackwell, Singapure.

Mc Gavin, M. D., Zachary, F. J. (2007). *Pathologic Basis of veterinary disease*. 4th ed. Mosby Elseivier, USA.

Minovich, F., Paludi, A., Cebey, P., Avefa, C. (2004). *Feline anasarca as a cause of dystocia*. http://www.veterinaria.org/revistas/redvet/n080804.html. (May 19, 2021).

Noakes, E. D., Parkinson, T. J., Gary, C. W. (2009). *Veterinary reproduction and obstetrics*. 9th ed. Saunders. England.

Osorio, B. J. C., Flores, V. L. M., Martinez, R. E. G., (2011). Fetal anasarca (hodrops toetalis) and congenital hypotrichosis in a dog: a case report. *XX National Congress of Veterinary Pathologists.* Puerto Vallarta, Jalisco, Mexico. pp. 21-25.

Patruck, L., Francis, W., Smith, H. (1998). *The veterinary consultation in 5 minutes, canine and feline.* Inter-americana, Argentina.

Peterson, M. E., Kutcher, M. A. (2011). *Small animal pediatrics. The first 12 mounths of the life.* Saunders. Chain.

Sucupira, M. C., Arcaute, M. R., Lacasta, D., Pérez, M. J., Ramos, J. J., Verde, M. T., Ferrer, L. M. (2011). *Congenital and hereditary diseases of the skin of* lambs. http://albéitar.portalveterinaria.com/noticia/8405/ART%C3%8DCLOS-RUMIANTES-ARCHIVO/enfermedadescong%C3%A9nitashereditarias-piel-corderos-i.html. (May 19, 2021).

Trigo, F. J, Valero, G. (2004). *Veterinary general pathology.* 4th ed. Department of Pathology Faculty of Veterinary Medicine and Zootechnics National Autonomous University of Mexico. México.

Trigo, F. J. (2011). *Veterinary systemic pathology.* 5th ed. Mc Graw-Hill. Mexico.

Contents of Earlier Volumes

Advances in Animal Science and Zoology. Volume 19

Advances in Animal Science and Zoology. Volume 18

Advances in Animal Science and Zoology. Volume 17

Advances in Animal Science and Zoology. Volume 16

Advances in Animal Science and Zoology. Volume 15

Index

E

F

G

H

I

K

L

M

N

O

P

R